Foundations of Soft Logic

Moshe Klein • Oded Maimon

Foundations of Soft Logic

Springer

Moshe Klein
Faculty of Engineering
Tel Aviv University
Tel Aviv, Israel

Oded Maimon
Faculty of Engineering
Tel Aviv University
Tel Aviv, Israel

ISBN 978-3-031-58235-6 ISBN 978-3-031-58233-2 (eBook)
https://doi.org/10.1007/978-3-031-58233-2

This Springer imprint is published by the registered company Springer Nature Switzerland AG
The registered company address is: Gewerbestrasse 11, 6330 Cham, Switzerland

Preface

Dear reader, we present to you our original research from the past seven years. It is a groundbreaking new mathematical paradigm and system (which includes the invention of a new number) that enables a new approach to advancing the foundations of science and technology.

We develop new approaches to connecting subjective and objective points of view that do not rely solely on the classical deductive approach. There is no need to fear paradoxical situations because it is precisely in them that lies the key to a new understanding that is more connected to life itself.

We wrote this book with the understanding that, in our times, there is a great need to develop a new approach for advancing science in the face of new challenges, including consciousness research. Our experience in the fields of robotics, computer science, and artificial intelligence has led us to the conclusion that the real challenge today is not only to make machines intelligent but also to develop and refine intelligence in human beings.

We would like to acknowledge and thank our friends who were involved in the research and assisted us in bringing it to the state of a complete book: to Yale Landsberg for the conversations that helped develop the Soft coordinate system (Chap. 5), to Prof. Kevin Vixie for the important contribution in linking the Soft numbers to the Möbius strip, to Prof. Alla Shmukler for the significant assistance in developing Soft analysis (Chaps. 9 and 10), to Oren Fivel for the significant contribution in developing Soft probability and Soft analysis (Chaps. 12 and 14), and to Dr. Ron Hirschprung for developing the research and implementation in the field of privacy paradox (Chaps. 13 and 14). Thanks also to Avishay Galili for the connection of Soft logic to Salomon Maimon and the Theory of Infinitesimals (Chap. 2), to Omer Cantor for his contribution to the understanding of the connection between Soft Möbius function and the Riemann hypothesis (Chap. 11), to Sigal Kordova for the correlation to System thinking (Chap. 14), John Torday for the connection to biology and the origin of life (Chap. 14), and to Idan Sagiv for his professional editing.

We thank the Koret Foundation Grant for Smart Cities and Digital Living 2030, bestowed upon the universities of Stanford and Tel Aviv, and especially Prof. Nick Bambos and Prof. Irad Ben-Gal.

Tel Aviv, Israel

Moshe Klein
Oded Maimon

Contents

1 Introduction 1
- 1.1 The Essence of Soft Logic 1
- 1.2 Overview of the Book 3
- References 6

2 Philosophical and Mathematical Background 7
- 2.1 Kant's Approach to Epistemology 7
- 2.2 Salomon Maimon and the Theory of Infinitesimals 8
- 2.3 Introduction to Soft Logic 9
- 2.4 The Evolution of the Number Concept 11
- 2.5 The Extension of the Concept of Infinitesimals 13
- References 16

3 Axioms of Soft Logic 17
- 3.1 Basic Definitions 17
- 3.2 The Axioms of Soft Logic 18
- 3.3 Bridge Numbers 19
  - 3.3.1 The Set of Bridge Numbers Is an Abelian Group with Respect to Addition 19
  - 3.3.2 The Set of Bridge Numbers Is a Ring 20
  - 3.3.3 Bridge Numbers Are Almost a Field 20
  - 3.3.4 The Power of Bridge Numbers 21
  - 3.3.5 Roots of Bridge Numbers 22
- Reference 24

4 Soft Numbers 25
- 4.1 The Axiom of Noncommutativity 25
- 4.2 Definition of Soft Numbers 26
- 4.3 The Algebra of Soft Numbers 27
  - 4.3.1 The Operations of Addition and Multiplication in the Set of Soft Numbers 27

4.3.2 Two Interesting Examples of Subgroups of $SN^*(\neg 0)$ 29
4.4 The Powers and Roots of Soft Numbers 29
Reference 30

5 The Soft Coordinate System 31
5.1 The Development of the Soft Coordinate System 31
5.2 Presentation of Soft Numbers 34
5.3 The Connection to the Möbius Strip 36
Reference 37

6 Soft Logic and Calculus 39
6.1 The Connection to the Derivatives Function 39
6.2 Soft Functions 41
6.2.1 Fundamental Properties of Soft Functions 41
Reference 43

7 Soft Curves 45
7.1 Definition of Soft Curves 45
7.2 The Fundamental Equation for Soft Curves 47
7.3 Presentation of the Symmetry of Two Mutually Inverse Functions 54
7.4 Soft Curves and Areas 56
Reference 57

8 The Dynamics of Soft Logic 59
8.1 Dynamics with Soft Numbers 59
8.2 Examples of the Dynamic with Soft Numbers 61
8.3 Soft Dynamics Geometry 66
8.4 Some Remarks Concerning the Results 71
References 73

9 Soft Analysis 75
9.1 Definition of the Distance in Any Set of Elements 75
9.2 The Distance Between Two Soft Numbers 79
9.3 The Neighborhood ε in the Set of Soft Numbers 80

10 Limits of Soft Numbers 85
10.1 The Limit of an Infinite Sequence of Soft Numbers 85
10.2 The Limit of a Soft Function at a Point 88
10.3 The Properties of the Limits of Soft Functions 90

11 The Soft Möbius Function and the Riemann Hypothesis 93
11.1 The Definition of the Soft Möbius Function 93
11.2 The Sum of the Soft Möbius Function 95
11.3 Connection to the Riemann Hypothesis 97
Reference 98

12 Soft Probability and Entropy 99
12.1 Introduction to Soft Probability 99
12.2 Defining Soft Probability 100
12.3 Complements, Unions, Intersections, and Soft Conditional Probability 102
12.3.1 Complements, Unions, and Intersections 102
12.4 Soft Expectation, Variance, and Entropy 105
12.4.1 Soft Expectation and Variance 106
12.4.2 Soft Entropy 109
12.5 Decision Trees Based on Soft Mutual Information 110
References 111

13 Soft Logic and the Privacy Paradox 113
13.1 Introduction to the Privacy Paradox 113
13.2 Representing the Privacy Paradox Using Soft Logic 114
13.3 Applying Soft Logic to the Privacy Paradox 116
13.4 Results and Conclusions 118
References 119

14 Application Examples and Further Research 121
14.1 Adopting Soft Logic to Systems Thinking 121
14.1.1 Representation of Systems Thinking 124
14.1.2 Ordering 126
14.2 Soft Logic Analysis 126
14.2.1 Our Motivation 127
14.2.2 Limits of Soft Number Sequences 127
14.2.3 Limits of Soft Functions 128
14.2.4 Derivatives of Soft Functions 129
14.3 Extending the Pascal Triangle with Soft Numbers 130
14.4 Conclusions and Further Research 132
References 133

Appendices 135
A.1 Nonstandard Analysis 135
A.2 Dual Numbers 136
A.3 Fractals 137
A.4 The Riemann Hypothesis 139

List of Symbols, Signs, and Terms of Soft Logic

Symbol/Sign/Term	Definition
$\bar{\mathbf{0}}$	The special object of the infinitesimal type with the fundamental property $\bar{0}^2 = 0$, the generator of the zero axis
$a\bar{\mathbf{0}}$ $(a \in \mathbb{R})$	A zero axis number, a Soft zero
$-\bar{\mathbf{0}} = -1\bar{\mathbf{0}}$	The negative zero
$+\bar{\mathbf{0}} = 1\bar{\mathbf{0}}$	The positive zero
$\mathbf{0} = 0\bar{\mathbf{0}}$	The absolute zero
$\bar{0}$-axis, a zero axis, a zero line	A vertical line presenting a continuum of Soft zeros
$\bar{\mathbf{1}}$	The number 1 when it is used to present a real number as a multiple of 1
$b\bar{\mathbf{1}}$ $(b \in \mathbb{R})$	A real number b presented as a multiple of 1
$\bar{1}$ –axis, a real axis, a real line	A vertical line presenting a continuum of real numbers
$\perp$	The bridge sign
$a\bar{\mathbf{0}}\perp b\bar{\mathbf{1}}$ $(a, b \in \mathbb{R})$	A bridge number of the right type (a real number is to the right of the bridge sign)
$b\bar{\mathbf{1}}\perp a\bar{\mathbf{0}}$ $(a, b \in \mathbb{R})$	A bridge number of the left type (a real number is to the left of the bridge sign). We tend to omit writing the one axis, so the regular notation of soft numbers is of the form $a\bar{0}\dot{+}b$
BN	The set of all bridge numbers. $\{ a\bar{\mathbf{0}}\perp b\bar{\mathbf{1}}, b\bar{\mathbf{1}}\perp a\bar{\mathbf{0}} \ : a,b \in \mathbb{R} \}$
$a + b\varepsilon$ $(a, b \in \mathbb{R}, \varepsilon^2 = 0)$	A dual number
$\dot{+}$	Soft number sign
$a\bar{\mathbf{0}} \dot{+} b$ $(a, b \in \mathbb{R})$	A Soft number in the right appearance (a real number is on the right side of the Soft number sign)
$b \dot{+} a\bar{\mathbf{0}}$ $(a, b \in \mathbb{R})$	A Soft number in the left appearance (a real number is on the left side of the Soft number sign)
SN	The set of all Soft numbers $\{ a\bar{\mathbf{0}}\dot{+}b \ : a, b \in \mathbb{R} \}$
$c = X\bar{\mathbf{0}}\perp Y$	A bridge number of the right type in the pair of bridge numbers defining a Soft number $X\bar{0}\dot{+}Y$

(continued)

Symbol/Sign/Term	Definition
$c' = Y \perp X\bar{0}$	A bridge number of the left type in the pair of bridge numbers defining a Soft number $X\bar{0} \dotplus Y$
SNS	A Soft number strip with a central zero axis and two bounding, real axes
$A\ (-\infty < A < \infty)$	The height of a point on an SNS
$B\ (0 \leq B \leq 1)$	The width of a point on an SNS
$c = (1 - B)A\bar{0} \perp BA\bar{1}$	A bridge number of the right type corresponds to a point C that is located in the right part of an SNS with a height A and a width B
$c' = BA\bar{1} \perp (1 - B)A\bar{0}$	A bridge number of the left type corresponds to a point C' that is located in the left part of an SNS with a height A and a width B
$\mathbf{S}N(+)$	The group of Soft numbers under addition
$\mathbf{S}N(+, *)$	The ring of Soft numbers under the operations of addition and multiplication
$\mathbf{SN}^*(\neg\mathbf{0})$	The set of invertible Soft numbers
$\mathbf{SN}^*(\mathbf{1})$	The set $\{a\bar{0} \dotplus (\pm 1) : a \text{ is any real number}\}$
$\mathbf{SQ}^*(\neg\mathbf{0})$	The set $\{q\bar{0} \dotplus r : q, r \text{ are rational numbers}, r \neq 0\}$
$\bar{f} : \mathbf{SN} \rightarrow \mathbf{SN}$	The Soft function of a Soft variable, an extension of a differentiable real function $f(x)$
$\widehat{f} : \mathbf{R} \rightarrow \mathbf{SN}$	The Soft function of a real variable, enabling a graphic presentation by a curve on an SNS

About the Authors

Moshe Klein is a researcher in the Consciousness and Soft Logic research lab at Tel Aviv University headed by Oded Maimon. He is the founder of the Gan Adam project for the development of mathematical and scientific thinking among pre-schoolers, and has since instructed hundreds of kindergarten teachers throughout Israel. He is a lecturer at Tel Hai College.

Oded Maimon is the Oracle chair professor at Tel Aviv University, previously at MIT. Oded's research focuses on exploring and modeling natural and artificial intelligence, Soft logic, and consciousness. He has spent sabbaticals at BU (Boston) and BHU (Varanasi, India), and was invited to lecture and help direct leading universities around the world (including Stanford in the United States, Tsinghua in China, IIT in India, Waseda in Japan, Paris 6 in France, and UNAD in Columbia). Oded has published over 100 refereed papers in leading academic journals and 20 research books, and has made over 200 research presentations at international scientific conferences with many invited and keynote presentations.

Chapter 1
Introduction

Soft logic is a mathematical language that facilitates richer and more diverse situations than the regular logical distinction between right and wrong. For this purpose, we have developed a rigorous mathematical theory that includes the invention of a new type of number called a "Soft number." This theory will be the foundation for scientific development and inventions in various new directions.

1.1 The Essence of Soft Logic

Our times fundamentally require a new mathematical language that is Soft and dialogical, a language that is based on the interaction between human beings and the world. The ability of Soft logic to contain contradictory and opposing situations may assist in situations of human conflict and dispute. This language enables each party to the conflict to develop a new point of view that sees and contains the point of view of the other party. This will make it possible to discover the common denominator between the two parties and resolve the conflict or disagreement.

Life is richer and more varied and colorful than the two extremities of right and wrong. To use the color analogy, one can say that this mathematical language is more colorful; it has more colors and shades than just black and white. Regular mathematics avoids paradoxical situations that contain internal contradictions. On the other hand, Soft logic by definition combines logical and linear thinking with a type of thinking based on the mathematical exercise of dividing zero by itself, an exercise with infinitely correct results, or, in other words, the type of thinking in which a thing and its opposite can be simultaneously correct.

Soft logic is based on a new perspective on the number zero, which was invented in India in the seventh century. This invention is relatively modern, compared to the invention of all the natural numbers tens of thousands of years ago. The invention of the zero was revolutionary since it gave a special sign to an amount that is nothing at all. This invention enabled the development of the decimal system of writing

M. Klein, O. Maimon, *Foundations of Soft Logic*,
https://doi.org/10.1007/978-3-031-58233-2_1

numbers, which was then used to write large numbers with a relatively small amount of digits. Soft logic expands the effect of the location of the number zero as it assumes a continuum of zeros rather than a discrete location of zeroes.

The mathematician Leibnitz developed the binary system based on two numbers only: one and zero; later it became the mathematical infrastructure for the development of the computer. Leibnitz himself strove to develop a new mathematical language that would be at the same time rational and softer than the dichotomy of right and wrong; the term "Soft logic" was inspired by him.

The novel distinction of Soft logic is that zero is not, in fact, nothing. To describe nothing, a blank sheet of paper would suffice. The very act of writing down the number zero shows that there is something that can be distinguished. This leads to the possibility that zero is both negative and positive. A similar distinction already exists in differential and integral calculus, which refers to the direction in which a series of points tends to zero. Contrary to this, Soft logic distinguishes between the various multiples of the number zero (Klein and Maimon 2021; Fivel et al. 2023; Hirschprung et al. 2023).

Soft logic has a zero axis, upon which lie the various multiples of the number zero. Soft logic offers a new set of coordinates that is non-Cartesian. This system has a zero axis facing the axis of real numbers. The Soft system of coordinates includes a twist and therefore constitutes a simple model for an infinite Möbius strip. Locally, a Möbius strip has two sides, but globally, it has only one side. This paradox is demonstrated and realized geometrically with Soft logic.

Using the zero axis, we define a new type of number called "Soft numbers." The following operations can be applied to them: addition, subtraction, multiplication, division, exponentiation, and extraction of roots. Soft numbers have an algebraic structure very similar to the structure of an algebraic field.

Complex numbers, which include the root of minus one ($i = \sqrt{-1}$), were invented 500 years ago by the Italian mathematician, Gerolamo Cardano. The mathematicians of his time found it very difficult to accept this concept, but after 400 years, it was put into extensive use in the mathematical development of the theory of electricity and alternate current.

Without the invention of complex numbers, the Internet could not have been invented. Complex numbers are also useful for the description of periodic waves in optical physics as well as quantum theory. In Soft logic, instead of the imaginary number i, we use the number $\overline{0}$. The laws of addition for Soft numbers are similar to those for complex numbers, but the laws of multiplication are different. Eventually, a use was found for complex numbers; similarly, we aim to develop and discover scientific and technological applications for Soft logic.

The relationship between mathematics and physics is one of the most challenging issues facing science today. It is one of the open problems (Problem no. 6) posed by the mathematician David Hilbert at the International Conference of Mathematics that took place in Paris in the year 1900. In 1905, Albert Einstein published the special theory of relativity, whose main idea is the constancy of the speed of light in a vacuum. Two sources of light that are moving toward each other do so at the relative velocity of the speed of light rather than at twice the speed of light. Ostensibly, this defies common sense. To describe the relative velocities of two bodies moving toward each other at certain velocities, Einstein used the mathematical transformations developed by the Dutch physicist, Hendrik Lorentz.

Einstein vigorously opposed quantum theory because he thought that it contradicted the principle of the limit of the speed of light. However, studies conducted in recent years have shown that there is a phenomenon in which particles that are naturally located at a distance from each other affect each other immediately, which seems to violate the speed of light limit. We postulate that the Soft coordinate system may be suitable for describing this immediate-effect phenomenon since, in the zero coordinate system, the distance between any two points is zero.

Soft logic corresponds with philosophical worldviews in which there is no separation between the phenomena experienced by an observer and the world itself; rather, they are intertwined. External reality affects one's internal reality, and one's internal reality generates and affects the external world. The philosopher Immanuel Kant developed an epistemology that distinguishes between the world as it is and the world of phenomena. The Jewish philosopher Salomon Maimon improved upon Kant's theory and argued that the real world exists in our consciousness as infinitesimal potentials. This corresponds with the model of Soft numbers, in which we have shown that the zero axis is a tangible model of the infinitesimal world.

The French mathematician and physicist Blaise Pascal, who proved the existence of vacuums, underwent a spiritual experience at the age of 32, after which he decided to part ways with science and mathematics and dedicate himself to philosophy and theology. Soft logic will enable bridging the gap that appears to lie between science and spirituality. At high levels of consciousness, language can affect and create reality. This is the vision of Soft logic.

1.2 Overview of the Book

Chapter 2 establishes the mathematical philosophy of Soft logic. The philosopher Immanuel Kant developed an epistemology that was revolutionary for its time. He created a distinction between the world of phenomena that is accessible to our senses and the world as it is, which is beyond our consciousness and is inaccessible to us. Kant argued that we are unable to investigate and understand the world as it is; we are only capable of understanding the world that is reflected within us and that the categories of time and space are not objective characteristics of the world; rather, they are characteristics that are bestowed upon objects solely by human consciousness. One of Kant's most important critics was the Jewish philosopher Salomon Maimon. The chapter describes Maimon's theory of infinitesimals, according to which the world is embodied by human consciousness in infinitely small sizes. Soft logic is a mathematical language that creates a tangible model of the infinitesimal world by distinguishing between the various multiples of the number zero. The first chapter concludes with a review of the evolution of the various types of numbers—

natural, integers, rational and irrational, real and complex—as well as the development of Soft numbers.

Like every methodical mathematical theory, Soft logic relies on several axioms. Accordingly, Chap. 3 presents the first four axioms of Soft logic. The first states that it is possible to distinguish between the various multiples of the number zero. The second axiom states that it is possible to connect the multiples of the number zero in a simple manner. The third axiom states that the product of two points on the zero axis collapses into absolute zero. The fourth axiom states that there is a bridge connecting real numbers and numbers on the zero axis. These axioms are used to define bridge numbers that can be added, subtracted, multiplied, divided, exponentiated, and their roots extracted. We show that the set of bridge numbers constitutes a quasi-algebraic field.

Chapter 4 presents the definition of Soft numbers. For this purpose, we formulate a fifth axiom relating to the properties of noncommutativity. The order in which we operate in the world changes the result we obtain in the end. Using this axiom, we define Soft numbers, which have two distinct appearances: on the right and on the left. In other words, the basic definition of the number concept is ambiguous. Similar to bridge numbers, we define the algebra of Soft numbers, which also constitutes a quasi-algebraic field.

Chapter 5 describes the Soft numbers system of coordinates. We detail the evolution and development of the coordinate system, including the vanishing point, followed by a description of the zero axis, which parallels the real axis, and the two real axes that appear reversed. We describe how it is possible to match Soft numbers to points on the Soft system of coordinates. The reversal of the two real axes enables viewing the Soft system of coordinates as a simple description of points on a Möbius strip. This is one of the properties that permits seeing the Soft numbers system as an organic system.

Chapter 6 offers a discussion of the natural connection between Soft logic and differential and integral calculus. When one examines the power of a Soft number, the connection to derivatives of functions is immediately apparent. We extend this simple property to polynomials and then to analytic functions. Subsequently, we define Soft functions, which are an expansion of real functions, and we prove fundamental properties of Soft functions.

In Chap. 7 we define Soft curves, which describe real functions through the Soft system of coordinates, and we present several examples of Soft curves. Subsequently, we prove the fundamental equation of Soft curves, and we also prove that the symmetry of curves is described by a function and its inverse. Finally, we prove the relationship between Soft curves and calculate the rectangular areas created by them.

In Chap. 8, we define the dynamics of Soft functions, meaning what happens when a Soft function is repeatedly applied on an initial number. We are interested in the mathematical properties of the sequence of numbers created in this process, and we show the connection that is created between the dynamics of Soft functions and high-order derivatives. We present several examples of the dynamics of specific Soft functions. In conclusion, we describe geometrically the dynamics of Soft functions, as inspired by fractal and chaos theory.

Chapter 9 presents an introduction to Soft analysis. We define the distance between two Soft numbers and the absolute value of Soft numbers. The innovation lies in understanding that the distance between two points on the zero axis is a type of Soft number. Using these properties, we define the fundamental *neighborhood* of Soft numbers, and the unique typology of the set of Soft numbers and their mathematical properties.

Chapter 10 discusses the limits of Soft numbers. We define the concept of a finite and infinite limit of an infinite sequence of Soft numbers. We define divergent and convergent sequences of Soft numbers, bounded and unbounded sequences, the limit of a Soft function at a point, and the basic properties of limits in the region of Soft numbers.

Chapter 11 extends the Möbius function of number theory to the Soft Möbius function. The regular Möbius function, which is related to prime factorization, accepts three values: 1, 0, and −1. Using Soft logic, we extend the function to the following five values: −1, −0, 0, +0, and 1. We prove the basic properties of the Soft Möbius function, and then we discuss the properties of the Soft Mertens function, which is the aggregate sum of the values of a Soft Möbius function. We show that the relationship between the properties of the Mertens function and the Riemann hypothesis is correct also when applied to the relationship between the Soft Mertens function and the Riemann hypothesis. We present the results of a computer simulation with a Soft Mertens function for numbers up to 10 million. We conclude the chapter with a description of the combinatory properties of the Soft Möbius function.

Chapter 12 discusses the development of probability theory through Soft logic. We open with the basic distinction that in ordinary probability theory, the probability of a specific event expanding is zero. Using Soft numbers, we suggest a mathematical model according to which the probability of a specific event expanding is a Soft zero. We define Soft probability and discuss complementary events, unions, intersections, and conditional Soft probability. Subsequently, we define the concepts of expected value, variance, and entropy in Soft probability. We conclude with a possible implementation of Soft information to the development of decision trees.

Chapter 13 describes a mathematical model of the privacy paradox using Soft numbers. The privacy paradox refers to the discrepancy between how people define the level of privacy they maintain on the Internet and their actual behavior. We define a mathematical model that creates a Soft number that is a product of the difference between these two characteristics. We describe a study on the privacy paradox conducted among a group of people and present the data graphically. In doing so, we indicate a possible direction for implementing Soft logic in the cyber world.

Chapter 14 presents conclusions regarding Soft logic and points out the possible applications and suggestions for future research.

References

Fivel, O., Klein, M., Maimon, O.: Soft decision trees. In: Rokach, L., Maimon, O., Shmueli, E. (eds.) Machine Learning for Data Science Handbook, pp. 143–170. Springer (2023)

Hirschprung, R.S., Kordova, S., Klein, M., Maimon, O.: Representation and assessment of systems thinking competencies through Soft logic. IEEE Access. **11**, 141547–141558 (2023). https://doi.org/10.1109/ACCESS.2023.3342131

Klein, M., Maimon, O.: Fundamentals of Soft logic. New Math. Nat. Comput. **17**(3), 703–737 (2021). https://doi.org/10.1142/S1793005721500356

Chapter 2
Philosophical and Mathematical Background

This chapter establishes the philosophical-mathematical foundation of Soft logic. We briefly describe Immanuel Kant's epistemology (Kant 1838), which was revolutionary for its time, especially for his critical distinction between the world of phenomena that our senses perceive and the world as it is in itself, the things as they stand outside the faculties of our consciousness. This distinction in Kant's epistemological theory entails difficulties which were recognized by Salomon Maimon's critique of Kant in his "Essay on Transcendental Philosophy" (Maimon 1790/2010). Maimon was a Jewish philosopher who developed the theory of metaphysical infinitesimals, arguing that the world is embodied by human consciousness in infinitely small sizes. Furthermore, we will present a tangible model of the infinitesimal world by distinguishing between the various multiples of the number zero. We conclude this chapter with a survey of the evolution of numbers and the development of the world of infinitesimals in mathematics (Ehrlich 2018).

2.1 Kant's Approach to Epistemology

We describe the philosophical background that inspired us and that helps explain the mathematical theory of "Soft logic," a language through which we aim to bridge two worlds: the infinitesimal-real world (which is beyond the understanding of humans, according to some philosophers) and the inner world of phenomena, appearances, representations, and processes, as perceived by our reasoning and senses. We do this by defining a new number called a "Soft number" (SN), which is composed of multiples of zero and a part that is real (the real numbers $\mathbb{R}$, as we know them). This yields two types of representation in an integrated model. Two of the important philosophical foundations for this are Immanuel Kant's epistemology and Salomon Maimon's doctrine of differentials, meaning his approach to infinitely small sizes.

Kant (1724–1804) argued that human consciousness has access only to the "phenomena," which are the world of sensible objects in time and space (Kant 1838).

M. Klein, O. Maimon, *Foundations of Soft Logic*,
https://doi.org/10.1007/978-3-031-58233-2_2

According to some interpretations of Kant's theory, the object about which the phenomenon revolved are the "things in themselves" or "noumena," a realm that is inaccessible to human consciousness. Thus, it actually turns out that the phenomenal objects stand in some unknown and undefined relation to the unknown world – the world of things in themselves. Despite their inaccessibility, consciousness must assume their existence (Kant did not use this term directly). Here, an irreconcilable gap arises in Kant's theory: What is the essence of the relationship between the things in themselves and the phenomena?

Kant's approach to epistemology was revolutionary because he posited the structure of human thinking as a central factor in the human perception of the world. Our knowledge of external objects is not founded on conclusions based on those objects; rather, Kant argued, there are synthetic a priori intuitions, namely, space and time—which alone enable us to be conscious of objects—and the medium, which enables them. Kant also contends that there are categories of reasoning that human consciousness uses to make judgments about the impressions it receives through space and time. Kant maintains that mathematics is synthetic a priori knowledge derived from the same source as the intuitions of space and time; it is a priori because it does not rely on experience, and it is synthetic because it does require a reconstruction through the act of intuition. Kant also holds that the source of our potential in mathematics is the same as that for formalizing objects in time and space.

2.2 Salomon Maimon and the Theory of Infinitesimals

One of the critics of Kant's epistemology in his time was Salomon Maimon. In his *Essay on Transcendental Philosophy*, Maimon analyzes, criticizes, and develops a new theory of Kant's weakest point, namely, "the things in themselves" (Maimon 1790/2010). He argued that a noumenon itself is an infinitesimal world of infinitely small-size relationships that are mathematical derivatives of pre-experiential processes of human consciousness. Every object, before it is perceived by human awareness, exists transcendentally as an infinitesimal size, having no quantitative size but only a qualitative one. The a-posteriori sensory images that is consciously observed, yield form the action of imagination and the intellect on the differentials.

> Considered in itself as a quality, every sensible representation must be abstracted from all quantity, whether extensive or intensive. For example, the representation of the color red must be thought of without any finite extension, although not as mathematical but rather as a physical point or as the differential of an extension. (Maimon 1790/2010, 19)

Salomon Maimon argued that there are two types of infinitely small sizes. The first arises when a series of sizes or numbers becomes smaller and tends toward zero without ever reaching zero. This is called a "potential infinitesimal". The second infinitely small size occurs when a quantity of nothing changes, and this is called an "actual infinitesimal."

In conclusion, it appears that one of the most important issues in epistemology is understanding the essence of the relationship between the world of things as they are and the phenomenological world. The nature of consciousness (which holds both aspects: the differential and the phenomena) is that we are compelled to assume that there is some kind of relationship between these two worlds. According to Kant, mathematics is the result of a synthesis between internal processes taking place in one's consciousness, along with observations of the world. In other words, mathematics essentially describes the possible affinity between the two worlds. According to Maimon, the phenomena of objects are the result of a mechanism of infinitesimal differentials that exist in human consciousness.

2.3 Introduction to Soft Logic

Soft logic is a new language of mathematics that facilitates and rigorously describes the two worlds and the bridges between them (Klein and Maimon 2019). The world of objects is mathematically represented as an axis of real numbers, while the world of inner consciousness is described by the zero axis, which is composed of multiples of the number zero. The relationship between them describes a new type of number called "Soft numbers," which are part real and part multiples of zero (expressing the ultimate, infinitely small sizes). One of the main theorems that result from Soft logic analysis is the relationship between the infinitesimal and the function itself. To make these ideas clear, we shall define them rigorously.

The development of Soft logic was also inspired by the research of Marcelo Dascal, who wrote about and described the work of the great mathematician and philosopher Gottfried Wilhelm Leibniz (Dascal 2008). Leibniz aspired to discover and develop a new mathematical language that would provide a softer logic and thereby overcome the limitations of the true/false dichotomy, which is a central concept in classical logic. According to Dascal, language is a tool for thought but also influences thinking. On many occasions, Leibniz argued that classical logic, which contains only the two states, true and false (conveniently represented by "1" and "0" in computational systems), is insufficient for grasping the full meaning of human reasoning and consciousness. However, he did not develop this idea any further and established only the initial thoughts of this theory.

The French mathematician, physicist, and philosopher, Blaise Pascal, proved the existence of the physical vacuum (Adamson 1995). We point out a rich mathematical world that exists in the abstract vacuum, which is infinitesimal (Ehrlich 2018) and infinite at the same time (two related concepts). Continuing Salomon Maimon's philosophical approach, we develop this world as an expansion of the number zero from a single point to a continuous line of zeros. We present an axiomatic theoretical development of the world without time and space (where entanglement processes and non-locality exist), which is tightly coupled with the regular phenomenological world (in a noncommutative way). The axioms themselves are inspired by the philosophical approaches introduced above. We also show the relevance of this work to the current dynamic reality in which we live (e.g., the swift addition of cyberspace to the infrastructure of our lives). One of the manifestations of this reality is the integration of the real world and the digital world, each having its own rules. An accurate description of these interwoven worlds requires the development of a

Fig. 2.1 Ramanujan

mathematical language that can bridge the two worlds, which ostensibly run in parallel but are in fact related and affect each other (though governed by different physical laws).

According to traditional mathematics, the expression 0/0 is undefined, although, in fact, any real number could represent this expression since $a * 0 = 0$ for all real numbers a. This observation opens up a new scope and area for investigation.

According to Pradip K. Datta, the special vision of the two zeros quotient was among Ramanujan's unique capabilities (Datta 2013). The Indian mathematician Ramanujan (Fig. 2.1), discovered almost 4000 original mathematical formulas. He did not prove the formulas that he discovered; he simply sketched them, as if creating a drawing. When asked how he knew that they were correct, he answered that the Indian goddess Namakel whispered them into his ear. Ramanujan is undoubtedly a mathematician who inspired an intuitive way of thinking with direct experience. Can we learn from his way?

The exercise of dividing zero by zero (0/0) demonstrates that there is a space of numbers in which everything is possible simultaneously (since there is no time and space, but there are distinct events and operators). Soft logic describes a mathematical language that bridges the real world and the symbolic world. The constant and ongoing distinction between the real and symbolic worlds creates "two eyes" that make it possible to navigate from a starting point and arrive at the desired solution or new formula.

The familiar Möbius strip is one geometric model that results directly from our theory and is suitable for describing the mutual influence of the two different worlds. The real world is on one side of the strip, while the world of information is on the other. Locally, the strip has two sides, but globally, it has only one (the unity of

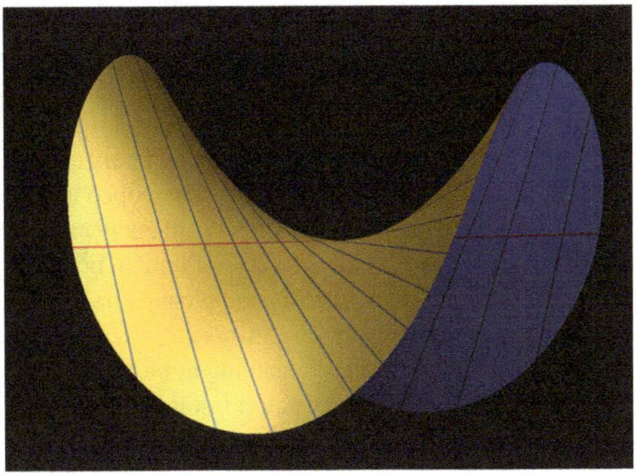

Fig. 2.2 Blowup

differences). Similarly, Soft logic naturally encompasses opposites and paradoxes (Starostin and Heijden 2007).

The mathematician Felix Klein developed a geometric-algebraic model called "blowup" (Álvarez et al. 2011) (Fig. 2.2). By blowing up the point of origin of some coordinate system, a Möbius strip is formed. To the point of origin, we add all of the possible revolutions around it, for a full 360°. A circle is thus created around the point of origin, which is connected by lines to each point on this circle. After turning the circle by 180 degrees, it returns to itself but in the opposite direction. When this procedure is conducted in three dimensions, reversing the direction upon itself by 180°, a twist of the Möbius strip is created.

In Soft logic, we blow the zero point by distinguishing between the various multiples of zero and creating a zero axis. Using five axioms, we define Soft numbers and develop their algebra, which we prove almost creates a mathematical field (omitting the mysterious zero axis, where there is no inverse). A field is a set of elements that are closed under two operations, addition and multiplication. These operations satisfy certain axioms. In a field, every element has an inverse under multiplication, despite the zero element. In this chapter, we briefly describe the Soft coordinate system, the algebra of Soft numbers, the relationship with differential and integral calculus, the expansion of Soft functions, and their presentation on Soft curves. To conclude, we present several practical avenues for future research using Soft logic. The full theory is presented in several references (Klein and Maimon 2016a, b, 2019, 2020, 2021).

2.4 The Evolution of the Number Concept

Throughout history, mathematicians have extended the "number" concept several times. It appears that every step in this evolution was the result of a need to cope with unsolvable problems, which required forging a new path forward. Each time, the new set was defined by the previous one, enabling the definition of new, fruitful mathematical concepts, thereby solving what could not previously be solved. This review is intended to provide a broad perspective for understanding the emergence of Soft numbers, which are another extension of the number concept.

Prehistoric man used natural prime numbers (1, 2, 3, 4 ... etc.), perhaps with his fingers or with stones. Numbers were developed in order to count objects. In Ancient Greece, it was understood that there were many natural numbers. According to Plato, natural numbers are a characteristic of human beings that sets them apart from other animals. The German-Jewish mathematician, Leopold Kronecker, said that God invented natural numbers and man invented the rest. In Archimedes' time, it was understood that a "largest" natural number does not exist because each one is followed by another, larger natural number. In other words, there is an infinite set of natural numbers, which today we designate by the letter N.

The next step in the evolution of the number concept was taken in order to provide a solution to equations of the form $x + m = n$ for all the natural numbers m, n (e.g., $x + 6 = 4$). For this purpose, numbers counter to natural numbers were invented in China; they are called negative integers. The number zero was invented in India in the seventh century to address the problem of giving a name to the number of elements in an empty set. The number zero has an additional, important role in representing real numbers in the decimal system.

The need to solve equations of the form $ax = b$ (e.g., $3x = 5$), which cannot be solved with whole numbers, brought about the invention of rational numbers, which are the quotient of two whole numbers (e.g., $x = \frac{5}{3}$). The set of rational numbers has been known for over 3000 years. In Pythagoras' time, irrational numbers were invented when it became clear that rational numbers do not entirely satisfy the need to measure the length of a segment of a line using another segment. One example of an irrational number is the solution to the equation $x = \sqrt{2}\, x^2 = 2$, which cannot be expressed as a quotient of integers.

The set of irrational numbers, together with the set of rational numbers, constitutes the set of real numbers. Presented geometrically, real numbers are all points on a line that is called the real axis, which is denoted by $\mathbb{R}$. Algebraic numbers are real numbers that are solutions to polynomials with integer coefficients (e.g., $\sqrt{2}$). Transcendental numbers are real numbers that are not algebraic, such as e, π.

One important problem could not be solved with real numbers, namely, equations such as $x^2 + 1 = 0$. The desire to solve this problem brought about the discovery of a new type of number in the sixteenth century. Called complex numbers, they have the form $a + bi$, where $i = \sqrt{-1}$. In their geometric presentation, complex numbers fill a plane that is called the Gauss plane, which has two axes perpendicular to each other: a real, horizontal axis, and an imaginary vertical axis. Further extensions of the set of numbers include the quaternions, which were invented by William Hamilton, and nonstandard numbers, invented by the Jewish mathematician Abraham Robinson; both are similar in essence to the extension of Soft numbers that we are suggesting.

Soft logic suggests further development of the number concept by understanding the various multiples of the number zero. The mathematical exercise of dividing zero by itself is considered undefined, but in fact, the multiplication of every real number

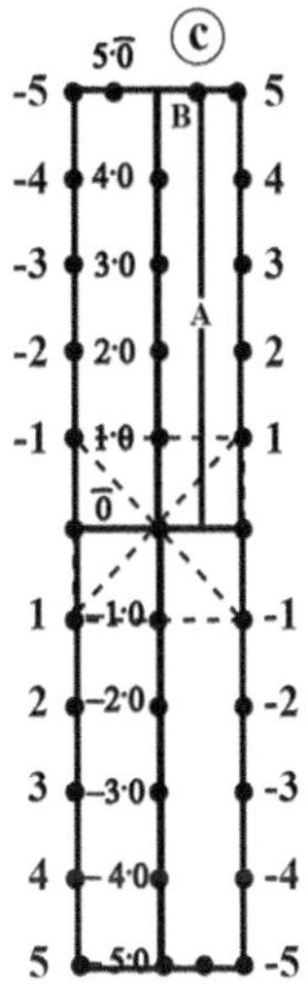

Fig. 2.3 The Soft coordinate system

by a denominator of zero results in a number that has zero as a numerator. Therefore, the result of the exercise of dividing zero by itself can be any real number. Since the denominator equals zero, we suggest the following solution: $\frac{0}{0} = a \cdot \bar{0}$, where $\bar{0}$ is the extension of the number zero to the zero axis. Soft numbers have the form $a\bar{0}\dot{+}b\bar{1}$, where $\bar{1}$ is the one axis of the real numbers. We tend to omit writing the one axis, so the regular notation of Soft numbers is of the form $a\bar{0}\dot{+}b$.

Soft logic is established as a mathematical theory through the use of axioms. Soft logic defines a new, non-Cartesian coordinate system (Fig. 2.3). There are two real axes facing each other and a zero axis in the center. Every Soft number can be described as a point within the Soft system of coordinates. Twisting the two real axes enables seeing Soft numbers as points on a Möbius strip. The result is the ability to use Soft numbers as a description that combines internal and external reality. Soft logic is an organic language that lives on the Möbius strip and its extension of the Klein bottle (Rapaport 2013) (Fig. 2.4).

2.5 The Extension of the Concept of Infinitesimals

An infinitesimal is a mathematical size that is closer to zero than any other real number. It does not exist in the real system of numbers because, for every real positive number, there is always another real number smaller than it. Infinitesimals exist in other number systems that have objects of an infinitely small or infinitely large size. Seemingly, the existence of infinitesimals contradicts common sense (Bell 2008) (Fig. 2.5).

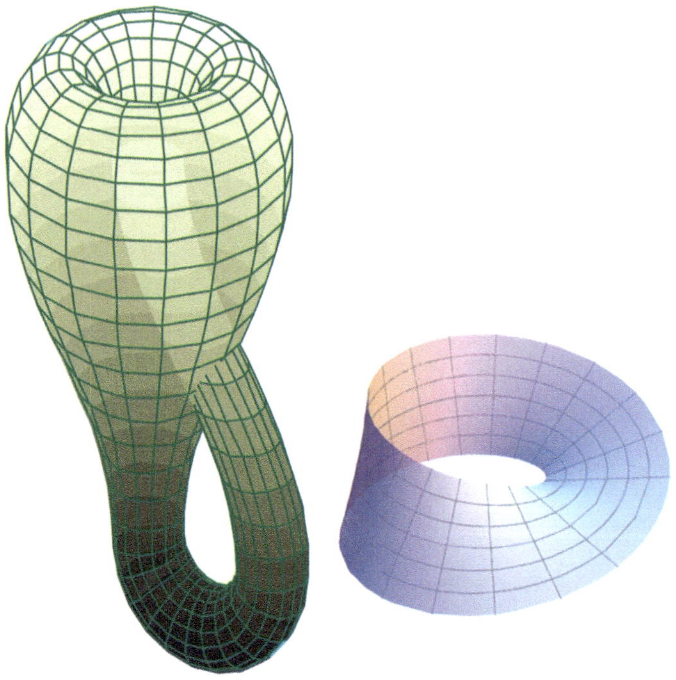

Fig. 2.4 A Klein bottle and a Möbius strip

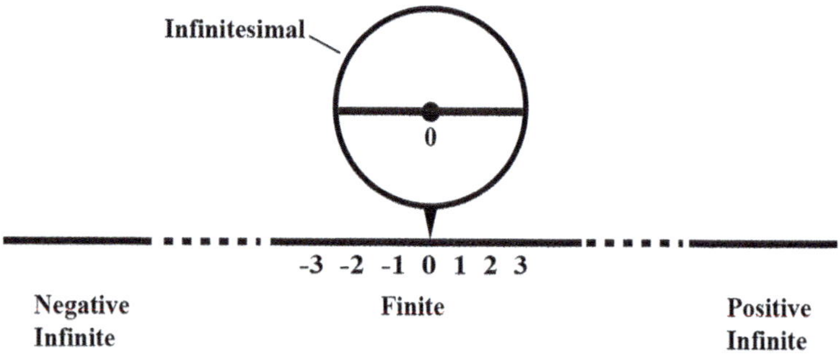

Fig. 2.5 Infinitesimals

The first to engage with infinitesimals and use them was the Greek mathematician Archimedes (287–212 BCE). He noted two important properties of real numbers:

1. By adding the number 1 to itself enough times, it is possible to exceed any real positive number x. In other words, $1 + 1 + 1 + \ldots + 1 > x$.
2. By adding a positive number x to itself an infinite number of times, no matter how small x is, it is possible to exceed 1. In other words, there exists

$$x + x + \ldots + x > 1$$

Therefore, every field of numbers that has these two properties is called an Archimedean field. Using infinitesimal methods, Archimedes discovered the formula for calculating the volume of three-dimensional objects, such as pyramids and cones.

The English mathematician John Wallis (1616–1703) introduced the expression $\frac{1}{\infty}$ in 1655 in his book on the cross sections of a cone. This expression describes infinitesimals as the inverse of infinity. Various mathematicians in medieval times and the modern age used the idea of infinitesimals. The Italian mathematician Bonaventura Cavalieri (1598–1647) formulated and invented the principle for calculating the area and volume of objects whose cross section is equal in length or area, now known as the Cavalieri principle. The French mathematician Pierre de Fermat (1607–1665) used infinitesimals in his proof that the slope of a smooth function at its maximum or minimum point is zero.

Two important mathematicians—the English mathematician Isaac Newton (1642–1727) and the German mathematician Gottfried Wilhelm Leibniz (1646–1716)—independently developed differential and integral calculus, which is based on the use of infinitesimal sizes. One of the prominent applications of this calculus is in the field of mechanics, in the calculation of the velocity and acceleration of bodies moving in space. This calculus was part of the scientific revolution begun by the Polish astronomer and mathematician Nicolaus Copernicus (1473–1543), the German astronomer and mathematician Johannes Kepler (1571–1630), and the Italian astronomer and mathematician Galileo Galilei (1584–1642).

One of the greatest critiques of the development of the calculus of infinitesimals was formulated by the Irish philosopher and bishop, George Berkeley (1685–1753), who argued that this entire theory is founded on contradictory assumptions: on the one hand, one divides by an infinitesimal because it is greater than zero, but then it is replaced by zero in order to obtain the final result of the calculation.

Between the seventeenth century and the middle of the nineteenth century, the methods of differential and integral calculus were widely used in the West, but it was obvious to everyone that the foundation of this theory was controversial. In light of this understanding, the fundamental concepts of differential and integral calculus were formalized in the nineteenth century by the French mathematician Augustin-Louis Cauchy (1789–1857) and the German mathematician Karl Weierstrass (1815–1897). They did this using the concept of limits, through which they defined other concepts such as derivatives, integrals, and the continuity of a function. In this way, the entire theory became established and accepted, from both the practical and philosophical points of view.

In the middle of the twentieth century, the idea of the existence of infinitesimal sizes was revived. The Israeli, Anglo-American mathematician, Abraham Robinson (1974), developed his nonstandard analysis, in which he provided a mathematical-logical foundation for the concept of infinitesimals and proved their existence using axioms from set theory. Although his theory was not constructive, it was accepted by many mathematicians and was widely used, with many varied developments. Another method for representing infinitesimals was developed using algebraic models, such as the dual numbers developed by the English mathematician (Clifford 1873), in which there exists an element $\varepsilon \neq 0$ that fulfills $\varepsilon^2 = 0$.

Soft logic expands the theory of infinitesimals through a new idea of extending the number zero along a continuous line that contains the number zero as well as various null sizes that are different from zero. The main innovation of Soft logic lies in the characterization of the concept of infinitesimals through the existence of the zero axis. We distinguish between the various multiples of a special number called "Soft zero," which is denoted by $\overline{0}$. Infinitesimal sizes are of the form $a\overline{0}$, where a is some real number. Based on the space created by the zero axis, we define a space of new numbers of the form $a\overline{0}\dot{+}b$, where b is a real number. These numbers are called Soft numbers.

References

Adamson, D.: Blaise Pascal: Mathematician, Physicist, and Thinker About God. Palgrave Macmillan, Basingstoke (1995)

Álvarez, M.J., Ferragut, A., Jarque, X.: A survey on the blow-up technique. Int. J. Bifurcation Chaos. **21**, 3103–3118 (2011)

Bell, J.A.: A Primer of Infinitesimal Analysis. Cambridge University Press (2008)

Clifford, W.: Preliminary sketch of bi-quaternions. Proc. Lond. Math. Soc. **4**, 381–395 (1873)

Dascal, M.: Leibniz's two-pronged dialectic. In: Dascal, M. (ed.) Leibniz: What Kind of Rationalist? Epistemology, and the Unity of Science, pp. 37–72. Springer, Berlin/Heidelberg/London (2008)

Datta, P.K.: National mathematics year: a tribute to Srinivasa Ramanujan. Sci. Cult. **79**(3–4), 158–162 (2013)

Ehrlich, P.: Contemporary infinitesimals theories of continua and their late 19th- and early 20th-century forerunners. (2018). https://doi.org/10.48550/arXiv.1808.03345

Kant, I.: The Critick of Pure Reason, Trans. Haywood F. William Pick, London (1838)

Klein, M., Maimon, O.: Soft logic and soft numbers. Pragmat. Cogn. **23**(3), 473–484 (2016a). https://doi.org/10.1075/pc.23.3.09kle

Klein, M., Maimon, O.: The mathematics of Soft logic. In: Paper presented at the IOP Second International Conference on Mechanical Engineering and Automation Science, Singapore, October 13–15 (2016b)

Klein, M., Maimon, O.: Axioms of Soft logic. P-Adic Numbers Ultrametric Anal. Appl. **11**(3), 205–215 (2019). https://doi.org/10.1134/S2070046619030038

Klein, M., Maimon, O.: The dynamics in the Soft numbers coordinate system. J. Adv. Math. **18**, 1–17 (2020). https://doi.org/10.24297/jam.v18i.8531

Klein, M., Maimon, O.: Fundamentals of Soft logic. New Math. Nat. Comput. **17**(3), 703–737 (2021). https://doi.org/10.1142/S1793005721500356

Maimon, S.: Essay on Transcendental Philosophy, Trans. Midgley N., Somers-Hall H., Welchman A., and Reglitz M. Continuum, London (1790/2010)

Rapaport, D.: Klein bottle logophysics: a unified principle for non-linear systems, cosmology, geophysics, biology, biomechanics, and perception. J. Phys. Conf. Ser. **437** (2013). https://doi.org/10.1088/1742-6596/437/1/012024

Robinson, A.: Non-Standard Analysis. Princeton University Press (1974)

Starostin, E., Van Der Heijden, G.: The shape of a Möbius strip. Nat. Mater. **6**, 563–567 (2007)

Chapter 3
Axioms of Soft Logic

This chapter presents the fundamental definitions of Soft logic. We reference the mathematical exercise of dividing zero by itself, which is usually considered unsolvable—even though every real number can be a solution. This notion led to the idea of the zero axis. We present the first four axioms of Soft logic, through which we define bridge numbers that can be added, subtracted, multiplied, divided, rooted, and exponentiated. We also show that the set of bridge numbers constitutes almost an algebraic field (Klein and Maimon 2019).

3.1 Basic Definitions

We start by stating the axioms upon which we develop the Soft logic theory. We define the zero axis, the 1 axis, the processes (basic addition and multiplication operations), and the relationship between the two worlds (bridge numbers, which are noncommutative, meaning that the influence of the axes on each other is nonsymmetric). There is also a specific point (the absolute zero) that will later emerge as a "rabbit hole," a mystery that does not obey any rule (including those that apply to the rest of the space). We will show that it serves as a torsion point, which allows the existence of some paradoxical properties to exist. We conclude this section with the presentation of a new number that captures both realities, as we perceive them, and the infinitesimal property of reality (the essential relation of reality).

The exercise of dividing zero by zero (0/0) demonstrates that there is a space of numbers in which everything is possible simultaneously (since there is no time and space, but there are distinct events and operators). Soft logic describes a mathematical language that bridges the real world and the symbolic world. The constant and ongoing distinction between the real and symbolic worlds creates "two eyes" that make it possible to navigate from a starting point and arrive at the desired solution or new formula.

M. Klein, O. Maimon, *Foundations of Soft Logic*,
https://doi.org/10.1007/978-3-031-58233-2_3

We assume the existence of a continuum of distinct multiples $a\bar{0}$, where a is any real number and $\bar{0}$ symbolizes an object that may be called a "Soft zero," while its multiples are also called "Soft zeros." We denote by $\bar{\mathbf{1}}$ the real number 1 when all other real numbers are conceived as its multiples. We also introduce the term absolute zero (bolded): $\mathbf{0} = 0\bar{0}$.

3.2 The Axioms of Soft Logic

Let a, b be any real numbers and $a\bar{0}, b\bar{0}$ two corresponding Soft zeros.

Axiom 3.1 (Distinction). If $a \neq b$ then $a\bar{0} \neq b\bar{0}$.

In Soft logic, we extend the zero from a point to a line (axis). This creates a distinction between different multiples of $\bar{\mathbf{0}}$ (per the logic of identity, since a and b are not the same). This axiom relies on the basic identity concept in philosophy. For example, if $a = 3$ and $b = 4$, $3\bar{0} \neq 4\bar{0}$ are two points on the zero axis. Both of them are zero, but they are different points.

From this axiom, we derive the first definition of Soft logic:

Definition 3.1 (Order). If $a < b$ then $a\bar{0} < b\bar{0}$.

Axiom 3.2 (Addition). $a\bar{0} + b\bar{0} = (a + b)\bar{0}$.

A continuum of zeros is a concrete model of the infinitesimal. In Soft logic, we develop a model that locates infinite numbers of zeros on a straight line. We define the addition of multiples of $\bar{0}$ as the addition of their corresponding real factors. For example, if a=3 and b=4, then the addition $3\bar{0} + 4\bar{0}$ on the zero axis creates another point, which is $7\bar{0}$. The $\bar{\mathbf{1}}$ axis behaves regularly: $a\bar{1} + b\bar{1} = (a + b)\bar{1}$.

Axiom 3.3 (Nullity). $a\bar{0} * b\bar{0} = 0$.

Numbers on the zero axis "collapse" under multiplication operations. The addition has significance and meaning, but the multiplication does not make any distinction whatsoever (these are the two extremes of wave interference for superposition operations: addition and nullity). For example, the above two points $(3\bar{0}, 4\bar{0}$ (result in an absolute-zero point if multiplied. This is like two infinitesimals that are squared and thus become smaller, approaching zero.

Multiplication represents a destruction process, while addition is an enhancement process. On the other hand

$$a\bar{1} * b\bar{1} = ab\bar{1},$$

$$a\bar{0} * b\bar{1} = ab\bar{0},$$

$$b\bar{1} * a\bar{0} = ba\bar{0}.$$

These equations describe the basic operations of a hybrid process.

Let us proceed and define a value that combines a value from the zero line ($\bar{0}$ axis) with a value from the real line ($\bar{1}$ axis) and denote it as $a\bar{0}\bot b\bar{1}$. Note that although the symbol between the two values reminds one of perpendicularity, it nevertheless has quite a different meaning: that of combining or "bridge making," and we call it the "bridge sign," which differs from the regular addition symbol (+).

Axiom 3.4 (Bridging). There exists a bridge between the zero axis and the real axis and vice versa, denoted by a pair of a bridge number and its mirror image on either side of the bridge sign: ($\boldsymbol{a\bar{0}\bot b\bar{1}}$, $\boldsymbol{b\bar{1}\bot a\bar{0}}$).

This is an existence axiom, stating that there is some special connection between two lines in the coordinate system. It implies that even two seemingly contradictory concepts (e.g., one concept lacks time and space, while the other has them) may have an affiliation and connection (e.g., our night dreams and daily reality). The next axiom is a property of this connection.

3.3 Bridge Numbers

Following the concept of the zero line, we define a new type of number called "bridge numbers." They have two components: multiples of $\bar{0}$ and multiples of $\bar{1}$. A bridge number is a number of the following form, where a and b are real numbers:

$$a\bar{0}\bot b\bar{1}$$

Note the new symbol $\bot$, which represents the bridging operation. It is not the same as a regular plus, and the whole expression has to be read as "$a\bar{0}$ bridge b." The set of all bridge numbers is $\mathrm{BN} = \{a\bar{0}\bot b\bar{1} : a, b \in \mathbb{R}\}$. For brevity, we ignore the presence of $\bar{1}$ and from here on out present Bridge numbers in the form

$$a\bar{0}\bot b$$

We do not yet place these numbers in the Soft coordinate system.

3.3.1 *The Set of Bridge Numbers Is an Abelian Group with Respect to Addition*

We define addition of bridge numbers according to Axiom 2 (Addition) of Soft logic:

$$\left(a\bar{0}\bot b\right) + \left(c\bar{0}\bot d\right) = \left(a\bar{0} + c\bar{0}\right)\bot(b+d) = (a+c)\bar{0}\bot(b+d)$$

The zero element of BN is $\left(0\bar{0}\bot 0\right)$ because $\left(a\bar{0}\bot b\right) + \left(0\bar{0}\bot 0\right) = a\bar{0}\bot b$.

The inverse element of $(a\bar{0}\perp \mathrm{b})$ in BN is $(-a\bar{0}\perp(-\mathrm{b}))$ because $(a\bar{0}\perp b)+(-a\bar{0}\perp(-b))=0\bar{0}\perp 0$.

The associative law is satisfied in BN:

$$(a\bar{0}\perp b)+((c\bar{0}\perp d)+(e\bar{0}\perp f))=$$
$$(a\bar{0}\perp b)+((c+e)\bar{0}\perp(d+f))=(a+c+e)\bar{0}\perp(b+d+f)=$$
$$((a\bar{0}\perp b)+(c\bar{0}\perp d))+(e\bar{0}\perp f)$$

Addition is also commutative:

$$(a\bar{0}\perp \mathrm{b})+(\mathrm{c}\bar{0}\perp \mathrm{d})=(\mathrm{c}\bar{0}+d)\perp(a\bar{0}\perp \mathrm{b})=(\mathrm{a}+\mathrm{c})\bar{0}\perp(\mathrm{b}+\mathrm{d}).$$

Thus, BN is a commutative group under the operation of addition.

3.3.2 The Set of Bridge Numbers Is a Ring

We define multiplication according to Axioms 2 and 3 of Soft logic:

$$(a\bar{0}\perp b)*(c\bar{0}\perp d)=(ad\bar{0}+bc\bar{0})\perp bd=(ad+bc)\bar{0}\perp bd$$

Under this definition, BN satisfies the distributive law:

$$(a\bar{0}\perp b)*((c\bar{0}\perp d)+(e\bar{0}\perp f))=$$
$$(a\bar{0}\perp b)*((c+e)\bar{0}\perp(d+f))=(ad+af+bc+be)\bar{0}\perp(bd+bf)=$$
$$(a\bar{0}\perp b)*(c\bar{0}\perp d)+(a\bar{0}\perp b)*(e\bar{0}\perp f)$$

Additionally, multiplication in BN satisfies the associativity law and has a neutral element $0\cdot\bar{0}\perp 1$ (clearly). Therefore, BN is a ring under addition and multiplication.

3.3.3 Bridge Numbers Are Almost a Field

Lemma 3.1 If $b\neq 0$, then $\frac{1}{a\bar{0}\perp b}=-\frac{a}{b^2}\bar{0}\perp\frac{1}{b}$.

Proof 3.1 The equality

$$\frac{1}{a\bar{0}\perp b} = x\bar{0}\perp y$$

means that

$$(a\bar{0}\perp b) * (x\bar{0}\perp y) = (ay + bx)\bar{0}\perp by = 1$$
$$by = 1$$
$$ay + bx = 0$$
$$y = \frac{1}{b}$$
$$x = -\frac{a}{b^2}$$

Therefore, BN is almost a field since all elements (excluding those on the zero line) have an inverse.

■

Lemma 3.2 Numbers on the $\bar{0}$ axis have no inverse.

Proof 3.2 For a Soft number $a\bar{0}\perp b$ on the $\bar{0}$ axis $b = 0$,

$$y = \frac{1}{b}$$
$$x = -\frac{a}{b^2}$$

are not defined.

■

3.3.4 *The Power of Bridge Numbers*

We can calculate the square of a Bridge number:

$$(a\bar{0}\perp b)^2 = \left(\overline{a0 * a0} + 2ab\overline{0}\right)\perp b^2 = 2ab\bar{0}\perp b^2$$
$$(a\bar{0}\perp b)^2 = 2ab\bar{0}\perp b^2$$

Calculation of powers $(a \cdot \bar{0}\perp b)^n$ where n is a positive integer is given by the following proposition:

Lemma 3.3 $\left(a\bar{0}\perp b\right)^n = nab^{n-1}\bar{0}\perp b^n$.

Proof 3.3

$$\left(a\bar{0}\perp b\right)^n = \left(a\bar{0}\perp b\right) * .. * \left(a\bar{0}\perp b\right)$$

In this multiplication, there are n terms of $\left(a\bar{0}\perp b\right)$. Every time we calculate the product, we need to choose between one of the terms, $\left(a\bar{0}\right)$ or (b). Any multiplication of two $\left(a\bar{0}\right)$ gives 0 and thus vanishes. Therefore $\left(a\bar{0}\right)$ must be multiplied only by $(n - 1)$ numbers b. There are n possibilities to choose an element $\left(a\bar{0}\right)$ that we multiply by (b^{n-1}). Thus, the component $nab^{n-1}\bar{0}$ before the bridge sign is obtained. To obtain the one after this sign, all n real numbers b are to be multiplied, which gives b^n. Therefore, $\left(a\bar{0}\perp b\right)^n = nab^{n-1}\bar{0}\perp b^n$. The statement can also be proved by induction for $n \geq 1$.

■

3.3.5 *Roots of Bridge Numbers*

We can calculate the square root of a bridge number:

$$\sqrt{a\bar{0}\perp b} = x\bar{0}\perp y$$

Lemma 3.4 If $b > 0$, then $\sqrt{a\bar{0}\perp b} = \pm\frac{a}{2\sqrt{b}}\bar{0}\perp \pm\sqrt{b}$.

Proof 3.4

$$\left(x\bar{0}\perp y\right)^2 = 2xy\bar{0}\perp y^2$$

$$2xy = a$$

$$y^2 = b$$

$$y = \pm\sqrt{b}$$

$$x = \pm\frac{a}{2\sqrt{b}}$$

For $b = 0$, $a \neq 0$ there is no solution (one cannot multiply a zero-axis number by itself and receive a zero-axis number, except when $a = 0$). Therefore, there are at most two square roots for every bridge number which is different from the absolute zero. We ascertain this by the following calculation:

$$\left(\frac{a}{2\sqrt{b}}\bar{0}\perp\sqrt{b}\right)^2 = 2\sqrt{b}\frac{a}{2\sqrt{b}}\bar{0}\perp\left(\sqrt{b}\right)^2 = a\bar{0}\perp b$$

$$\left(\frac{-a}{2\sqrt{b}}\bar{0}\perp-\sqrt{b}\right)^2 = 2\left(-\sqrt{b}\right)\frac{-a}{2\sqrt{b}}\bar{0}\perp\left(-\sqrt{b}\right)^2 = a\bar{0}\perp b$$

■

Lemma 3.5 The general n-th root of a bridge number is calculated as follows: For $b > 0$ or an odd n, $\sqrt[n]{a\bar{0}\perp b} = \frac{a}{nb^{\frac{n-1}{n}}}\bar{0}\perp\sqrt[n]{b}$, and for an even n also $\sqrt[n]{a\bar{0}\perp b} = -\frac{a}{nb^{\frac{n-1}{n}}}\bar{0}\perp-\sqrt[n]{b}$. For $b = 0$ and $a \neq 0$, there is no solution.

Proof 3.5

$$(x\bar{0}\perp y)^n = nxy^{n-1}\bar{0}\perp y^n$$

$$a\bar{0} + b = nxy^{n-1}\bar{0}\perp y^n$$

$$y^n = b$$

$$nxy^{n-1} = a$$

For an odd n (assuming b is nonzero),

$$y = \sqrt[n]{b}$$

$$x = \frac{a}{ny^{n-1}} = \frac{a}{nb^{\frac{n-1}{n}}}$$

For $b = 0$ and $a \neq 0$, there is no solution.
We can verify this by the following calculation:

$$\left(\frac{a}{nb^{\frac{n-1}{n}}}\bar{0}\perp\sqrt[n]{b}\right)^n = n\left(\sqrt[n]{b}\right)^{n-1}\frac{a}{nb^{\frac{n-1}{n}}}\bar{0}\perp\left(\sqrt[n]{b}\right)^n = a\bar{0}\perp b$$

For an even n and $b>0$, the calculation is similar to the one above, resulting in

$$y = \pm\sqrt[n]{b}$$

$$x = \frac{a}{ny^{n-1}} = \frac{\pm a}{nb^{\frac{n-1}{n}}}$$

We can verify the additional minus formula by the following calculation:

$$\left(\frac{-a}{nb^{\frac{n-1}{n}}}\bar{0}\perp - \sqrt[n]{b}\right)^n = n\left(-\sqrt[n]{b}\right)^{n-1}\frac{-a}{nb^{\frac{n-1}{n}}}\bar{0}\perp\left(-\sqrt[n]{b}\right)^n = a\bar{0}\perp b$$

■

We showed the richness of the algebra of BN: It is a commutative group, a ring, and almost a field, and the powers and roots have a particular nature.

Reference

Klein, M., Maimon, O.: Axioms of soft logic. Axioms of soft logic. P-Adic numbers ultrametric. Anal. Appl. **11**(3), 205–215 (2019). https://doi.org/10.1134/S2070046619030038

Chapter 4
Soft Numbers

In this chapter, we discuss the extension of the bridge numbers, mentioned in the previous chapter, to Soft numbers. For this purpose, we now add a fifth axiom to Soft logic, called "noncommutativity." When we engage in various activities in the world, the order in which we do them affects the final result. Let us use backgammon as an example: Assuming we rolled the dice and they came up 4 and 3, then sometimes we will be able to move three points and then four points, while we are unable to move three points and then four. This demonstrates the idea of noncommutativity (Klein and Maimon 2021).

4.1 The Axiom of Noncommutativity

The fifth axiom of Soft logic states that the order in which numbers are bridged is noncommutative. This axiom allows us to define Soft numbers, which are a combination of bridging on one side and bridging on the other side. Soft numbers are a type of superposition of the two nonsymmetric states of bridging. Following this definition, we develop the algebra of Soft numbers, which create a quasi-algebraic field. The algebraic properties of Soft numbers are identical to those of bridge numbers; the difference between them is manifested in the development of the Soft coordinate system.

Axiom 4.1 (Noncommutativity). $a\bar{0}\perp b\bar{1} \neq b\bar{1}\perp a\bar{0}$.

Note that the multiplication of a zero-axis number with a regular real number, as defined in Axiom 3.3, is commutative, but for the subsequent parts of the theory, the order of these numbers is important.

The two above combinations—a Soft zero and a real number with a bridge sign between them—will be identified by us as bridge numbers of two different types: a bridge number of a right type (where a real component is to the right of the bridge sign) and a bridge number of a left type (a real component is to the left of the bridge

M. Klein, O. Maimon, *Foundations of Soft Logic*,
https://doi.org/10.1007/978-3-031-58233-2_4

sign). By the noncommutativity Axiom 4.1, bridge numbers with the same components but of different types have different values. When dealing with two bridge numbers of the same type, the operations of addition and multiplication can be defined simply. However, to define such operations with two bridge numbers of different types is a serious challenge. For example, how to define the sum $(a\bar{0}\perp b\bar{1}) + (c\bar{1}\perp d\bar{0}) = ?$

We cannot apply any operations to the right and left appearance of bridge numbers; this is why we need to define the concept of Soft numbers.

4.2 Definition of Soft Numbers

Definition 4.1 A Soft number is a construction of the form $a\bar{0} \dot{+} b\bar{1} = \{ (a\bar{0})\perp(b\bar{1}); (b\bar{1})\perp(a\bar{0})\}$, where a, b are any real numbers.

Any Soft number in its denomination $a\bar{0}\dot{+}b\bar{1}$ has a component from the real line, or $\bar{1}$ line, and a component from the $\bar{0}$ line (later on, we will omit the line above 1, or even the number 1 itself, but the line over the zero is essential). Notice the new operation $\dot{+}$ that we define here. The dot above the plus sign implies that a Soft number has not one value, as a regular sum, but two distinct values, as presented in its definition. Every Soft number has a right appearance $a\bar{0}\dot{+}b$ and a left appearance $b\dot{+}a\bar{0}$. Is the content of a Soft number influenced by the change of the order of the components in its appearance? This matter is discussed below.

We present a Soft number in the following way:

$$a\bar{0}\dot{+}b = \{ c, c'\},$$

where

$$c = a\bar{0}\perp b,$$

$$c' = b\perp a\bar{0}.$$

According to the noncommutativity axiom, $a\bar{0}\perp b \neq b\perp a\bar{0}$, i.e., $c \neq c'$. However, as any set is by definition unordered, the pair $\{ c, c' \}$ is unordered, i.e., $\{ c, c' \} = \{c', c\}$.

Therefore $a\bar{0}\dot{+}b = b\dot{+}a\bar{0}$. By this equation, a Soft number in both its appearances, the right one and the left one, has the same two values.

In our theory, **SN** denotes the set of all Soft numbers:

$$\mathbf{SN} = \{a\bar{0}\dot{+}b : a, b \in \mathbb{R}\}.$$

It is important to remember that any Soft number in this set has two values, as each number consists of a bridge number $a\bar{0}\perp b$ and its mirror image $b\perp a\bar{0}$, which have different meanings by the noncommutativity axiom.

4.3 The Algebra of Soft Numbers

In this section, we develop the algebra of Soft numbers. We will show that it is possible to define a group **SN(+)** and a ring **SN(+**, x) with a set of Soft numbers and that this ring is almost a field. Furthermore, the set of elements of **SN(+**, x), which are invertible under multiplication, is a group that we will denote by **SN**$^*(\neg 0)$.

4.3.1 The Operations of Addition and Multiplication in the Set of Soft Numbers

We define the addition of two Soft numbers with the following rule:

$$\left(a\bar{0}\dot{+}b\right) + \left(c\bar{0}\dot{+}d\right) = (a+c)\bar{0}\dot{+}(b+d).$$

With this definition, the set **SN(+)** is a group under addition, because of the following:

(1) **SN** is closed under addition.
(2) **SN** has an identity (a neutral) element under addition, which is $0\times\overline{0}\dot{+}0$.
(3) Every element in **SN** has an inverse (an opposite) element under addition:

$$\left(a\bar{0}\dot{+}b\right) + \left(-\,a\bar{0}\dot{+}(-\,b)\right) = 0\bar{0}\dot{+}0.$$

(4) The associativity law is satisfied:

$$\left(\left(a\bar{0}\dot{+}b\right) + \left(c\bar{0}\dot{+}d\right)\right) + \left(e\bar{0}\dot{+}f\right) =$$
$$= \left(a\bar{0}\dot{+}b\right) + \left(\left(c\bar{0}\dot{+}d\right) + \left(e\bar{0}\dot{+}f\right)\right).$$

Proof of this statement is based on the addition definition and can be easily reproduced by the reader, so we do not display it here.

■

By the exposition above, **SN**(+) is a group. Moreover, it is an abelian, i.e., a commutative group, as the addition definition involves

$$\left(a\bar{0}\dot{+}b\right)+\left(c\bar{0}\dot{+}d\right)=\left(c\bar{0}\dot{+}d\right)+\left(a\bar{0}\dot{+}b\right).$$

We define the multiplication of Soft numbers by the following rule:

$$\left(a\bar{0}\dot{+}b\right)\times\left(c\bar{0}\dot{+}d\right)=(ad+bc)\bar{0}\dot{+}(bd).$$

This operation is commutative and satisfies the laws of associativity and distributivity. The formulation and verification of the first law are quite clear, and the second law is formulated and verified below.

The law of distributivity of Soft numbers:

$$\left(a\bar{0}\dot{+}b\right)\times\left(\left(c\bar{0}\dot{+}d\right)+\left(e\bar{0}\dot{+}f\right)\right)=$$

$$\left(a\bar{0}\dot{+}b\right)\times\left(c\bar{0}\dot{+}d\right)+\left(a\bar{0}\dot{+}b\right)\times\left(e\bar{0}\dot{+}f\right).$$

The proof of this statement is based on the definitions of addition and multiplication of Soft numbers as follows:

$$\left(a\bar{0}\dot{+}b\right)\times\left(\left(c\bar{0}\dot{+}d\right)+\left(e\bar{0}\dot{+}f\right)\right)=$$

$$\left(a\bar{0}\dot{+}b\right)\times\left((c+e)\bar{0}\dot{+}(d+f)\right)=$$

$$(ad+af+bc+be)\bar{0}\dot{+}(bd+bf)=$$

$$\left(a\bar{0}\dot{+}b\right)\times\left(c\bar{0}\dot{+}d\right)+\left(a\bar{0}\dot{+}b\right)\times\left(e\bar{0}\dot{+}f\right).$$

With these two operations + and x, Soft numbers create the Ring **SN**(**+**, x).

The Soft number $0\bar{0}\dot{+}1$ is a neutral element under multiplication in this set because for any of its elements $a\bar{0}\dot{+}b$, by the multiplication rule there is

$$\left(a\bar{0}\dot{+}b\right)\times\left(0\bar{0}\dot{+}1\right)=a\bar{0}+b.$$

Lemma 4.1 If $b\neq 0$, then a Soft number $a\bar{0}\dot{+}b$ has an inverse $\left(-\frac{a}{b^2}\bar{0}\dot{+}\frac{1}{b}\right)$.

Proof 4.1 To prove this statement, we have to ascertain that the product of both Soft numbers equals the neutral element of multiplication. By the rule of multiplication, we have

$$\left(a\bar{0}\dot{+}b\right)\times\left(-\frac{a}{b^2}\bar{0}\dot{+}\frac{1}{b}\right)=\left(\frac{a}{b}-\frac{a}{b}\right)\bar{0}\dot{+}\frac{b}{b}=0\bar{0}\dot{+}1.$$

The product is really the neutral element of multiplication, and the Lemma 4.1 statement is proved.

■

When a common notation for an inverse element is used, it may be written in the following form:

$$\left(a\bar{0}\dot{+}b\right)^{-1}=\left(-\frac{a}{b^2}\bar{0}\dot{+}\frac{1}{b}\right).$$

Let us observe the set of all Soft numbers $a\bar{0}\dot{+}b$, where $b\neq 0$. The above development demonstrates that this set forms a group under the operation of multiplication, which we denote **SN**$^*(\neg 0)$. Therefore **SN**$(+,x)$ is almost a field ("almost" because the inverse is undefined only when $b=0$).

4.3.2 Two Interesting Examples of Subgroups of SN(¬ 0)*

The set $\mathbf{SN}^*(\mathbf{1})=\left\{a\bar{0}\dot{+}(\pm 1)\text{ : a is any real number}\right\}$. The set is closed under multiplication because $\left(a\bar{0}\dot{+}(\pm 1)\right)\times\left(c\bar{0}\dot{+}\ (\pm 1)\right)=(\pm a\pm c)\bar{0}\dot{+}(\pm 1)$. (Note that the number 1 in the two factors may have opposite signs.) Every element in this set has an inverse: $\left(a\bar{0}\dot{+}(\pm 1)\right)^{-1}=-a\bar{0}\ \dot{+}(\pm\ 1)$, where the sign of 1 (+ or −) on both sides of "=" is meant to be the same.

The set $\mathbf{SQ}^*(\neg\mathbf{0}\,)=\left\{q\bar{0}\dot{+}r : q, r\ \textit{are rational numbers}, r\neq 0\ \right\}$. Since the multiplication and division of rational numbers (with a nonzero denominator) gives a rational number, this set is a subgroup of **SN**$^*(\neg 0)$.

4.4 The Powers and Roots of Soft Numbers

This section presents the properties of the roots and exponents of Soft numbers. The results are similar to the properties of bridge numbers, and therefore we present the results only without the proofs, which are identical to the proofs of the same theorems for bridge numbers.

Lemma 4.2 The n-th power of a Soft number is $\left(a\bar{0}\dot{+}b\right)^n=nab^{n-1}\bar{0}\dot{+}b^n$, where n is any natural number ($n=1,2,3,\ldots$). Now we wish to calculate the square root of a Soft number of the form $\sqrt{a\bar{0}\dot{+}b}=x\bar{0}\dot{+}y$.

Lemma 4.3 For $b > 0$ there is $\sqrt{a\bar{0}\dot{+}b} = \pm\frac{a}{2\sqrt{b}}\bar{0}\dot{+}\left(\pm\sqrt{b}\right)$, where $\sqrt{b}$ denotes an arithmetic (positive) square root of b and the signs in both components of the Soft root above are simultaneously + or simultaneously $-$. Obviously, for a Soft number $a\bar{0}\dot{+}b$ with $b < 0$, a square root does not exist. For $b = 0$, $a \neq 0$, there is no square root as well; it could only be a zero-axis number, and it is impossible to multiply a zero-axis number by itself and to receive a zero-axis number $a\bar{0}$ with $a \neq 0$.

If $a = b = 0$, i.e., a Soft number is an absolute zero, then its square roots are all Soft zeroes. Therefore, there are at most two square roots for every Soft number other than the absolute zero.

■

Lemma 4.4 The n-th root of a Soft number satisfies the following statements:

(1) $\sqrt[n]{a\bar{0}\dot{+}b} = \frac{a}{nb^{\frac{n-1}{n}}}\bar{0}\dot{+}\sqrt[n]{b}$ for $b \neq 0$ and an odd n.
(2) $\sqrt[n]{a\bar{0}\dot{+}b} = \pm\frac{a}{nb^{\frac{n-1}{n}}}\bar{0}\dot{+}\left(\pm\sqrt[n]{b}\right)$ for $b > 0$ and an even n.
(3) $\sqrt[n]{a\bar{0}\dot{+}b}$ does not exist for $b < 0$ and an even n and for $b = 0$ and $a \neq 0$.
(4) If $a = b = 0$, the root exists and can be any number on the zero axis.

Reference

Klein, M., Maimon, O.: Fundamentals of soft logic. New Math. Nat. Comput. **17**(3), 703–737 (2021). https://doi.org/10.1142/S1793005721500356

Chapter 5
The Soft Coordinate System

In this chapter, we proceed to define a new Soft coordinate system. This creation of a geometrical space allows the visualization of various processes. This system also enables us to present Soft numbers on an infinite strip and create the Möbius function, which shows the geometry of the space. The Möbius representation allows local antagonistic points to become the same globally, thus mathematically capturing one of the mysteries of consciousness, where differences may be transformed into unity (i.e., concepts of duality and non-duality can coexist, depending on the observer's context) (Klein and Maimon 2021).

5.1 The Development of the Soft Coordinate System

We suggest a new coordinate axis as presented in Fig. 5.1. It starts from 0 to 1 horizontally and then it turns 90° from 1 to infinity.

Lemma 5.1 There exists a one-to-one correspondence between the segment $(0, 1]$ and the segment $[1, \infty)$.

Proof 5.1 Consider the following function $f(x) = \frac{1}{x}$. This function creates a one-to-one correspondence between $(0, 1]$ and $[1, \infty)$.

■

Figure 5.1 contains an example of a line between $x = 1/2$ and $x = 2$. Also, one can draw such a line for all positive real xs other than 1, from x to $1/x$ as seen in Fig. 5.2.

Lemma 5.2 All the lines that connect x to $1/x$ (for all nonzero real x) intersect at a single point.

Proof 5.2 Let us observe the following drawing with a line, which *connects* x with $1/x$ as shown in Fig. 5.3.

$\Delta ABC \approx \Delta CDE$ as triangles with equal angles. Consequently,

M. Klein, O. Maimon, *Foundations of Soft Logic*,
https://doi.org/10.1007/978-3-031-58233-2_5

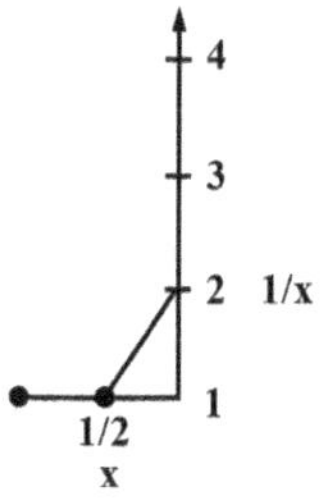

Fig. 5.1 The Soft coordinate axis

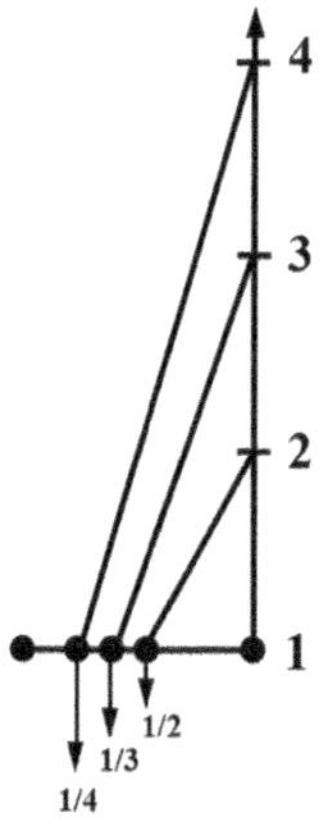

Fig. 5.2 The lines between x and $1/x$

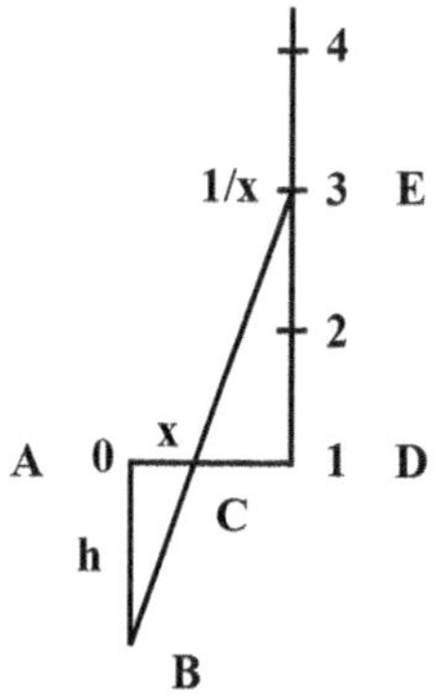

Fig. 5.3 The intersecting point

$$\frac{AC}{AB} = \frac{CD}{ED}$$

and therefore,

$$\frac{x}{h} = \frac{1-x}{\frac{1}{x}-1}$$

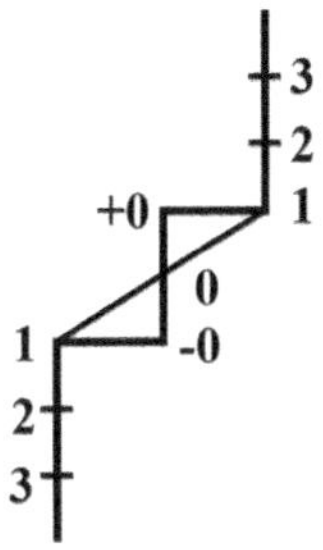

Fig. 5.4 The distinction between −0 and +0

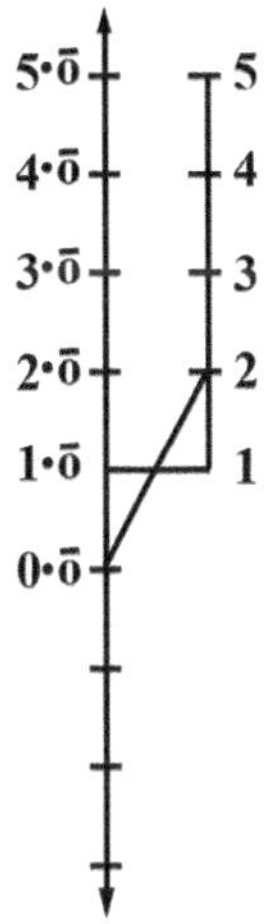

Fig. 5.5 The zero axis

and hence

$$h = 1.$$

This means that all the lines that connect x to $1/x$ intersect at the same point, which is located one unit below the zero.

■

The intersection point denotes the beginning of the Soft coordinate system. We call this point "absolute zero." The distance from absolute zero to $+\overline{\mathbf{0}}$ is one unit. We suggest extending this new coordinate axis symmetrically as shown in Fig. 5.4.

We now have three zeroes instead of one zero. One zero is opposite to the number −1, and it is not identical to the zero opposite the number +1. Hence, we suggest denoting the two different "zeroes" as $-\overline{\mathbf{0}}$ and $+\overline{\mathbf{0}}$. Furthermore, the discovery of the *zero line* enables us to find the locations of the multiples of Soft zero as shown in Fig. 5.5.

The zero line is developed according to the three axioms presented above. The following drawing (Fig. 5.6) presents the extended new coordinate system for positive and negative numbers with the multiples of Soft zero.

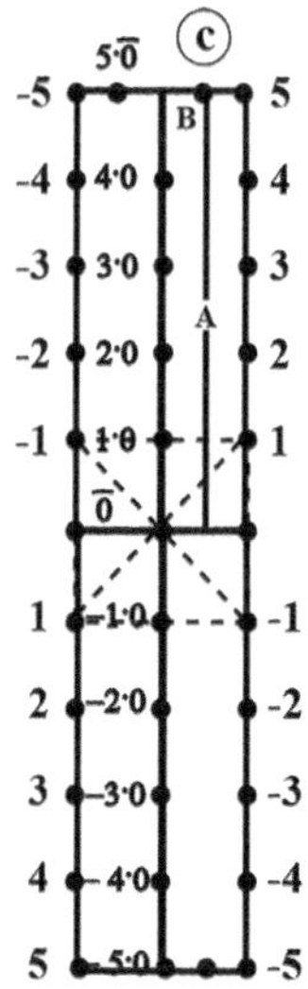

Fig. 5.6 The complete Soft coordinate system

5.2 Presentation of Soft Numbers

In Fig. 5.6, the middle vertical line is the zero axis that includes all the multiples of 0: $x\bar{0}$. The right-hand vertical line is the 1 axis that includes all the multiples of 1: $y\bar{1}$. The left-hand vertical line is the 1 axis that includes all the multiples of 1: $y\bar{1}$ but in the opposite direction. This coordinate system allows us to present all of the real numbers and all of the Soft zeros. We can now use this coordinate system for representing various Soft numbers. Given a Soft number $C = x\bar{0} \dot{+} y$, where x and y are any real numbers, we represent both elements of a Soft number as two points, on either side of a page. The point $x\bar{0} \perp y$ appears on one side of the page, and the point $y \perp x\bar{0}$ is the same point but on the other side of the page. Thus,

$$C = \{c, c'\} = \{\, x\bar{0} \perp y, y \perp x\bar{0}\}$$

$$c = x\bar{0} \perp y$$

is located on the front side of the page, while

$$c' = y \perp x\bar{0}$$

is located *on the back side of* the page.

As the infinite strip presented (partially) in Fig. 5.6 is intended for the presentation of Soft numbers, we call it a "Soft number strip" (SNS).

Definition 5.1 Let C be any point on the SNS. The height of point C is the vertical distance from C to the horizontal segment with the absolute zero at its center. A denotes the height with a sign. The width of point C is the horizontal distance from C to the zero line with a sign and is denoted by B.

The definitions above provide every point C on the SNS with two parameters, A and B, where $-\infty < A < \infty$, $-1 \leq B \leq 1$. The condition $A > 0$ is satisfied in the positive part of the SNS; $A < 0$, in its negative part, or correspondingly, above and below the horizontal segment containing the absolute zero, while on this segment $A = 0$.

For the second parameter, B, there is $B = 0$ on the zero axis, $B = 1$ on the right line bounding the SNS, $B = -1$ on the left line bounding the SNS, and otherwise $-1 < B < 1$.

Given $A \in \mathbb{R}$, $B \in [-1, 1]$, we define $C(A, B)$ as follows:

For $0 \leq B \leq 1$,

$c = (1 - B)A\bar{0} \perp BA\bar{1}$, located on the front side of the page.

$c' = BA\bar{1} \perp (1 - B)A\bar{0}$, located on the back side of the page.

$$C = x\bar{0} \dotplus \mathsf{y} = \{\mathsf{c}, \mathsf{c}'\}$$

For $-1 \leq B \leq 0$

$c = (1 + B)A\bar{0} \perp BA\bar{1}$, located on the front side of the page.

$c' = BA\bar{1} \perp (1 + B)A\bar{0}$, located on the back side of the page.

$$x\bar{0} \dotplus \mathsf{y} = \{\mathsf{c}, \mathsf{c}'\}$$

Note that if we use $|B|$, the two equations above are the same.

So, in brief, and in terms of the absolute value of B:

$c = (1 - |B|)A\bar{0} \perp BA\bar{1}$, located on the front side of the page,

$c' = BA\bar{1} \perp (1 - |B|)A\bar{0}$, located on the back side of the page, and therefore

$$x\bar{0} \dotplus \mathsf{y} = \{\mathsf{c}, \mathsf{c}'\}$$

Hence, by a coefficient comparison of the real part and the Soft part in

$$x = (1 - |B|)A,$$

$$y = BA$$

or equivalently, after solving the above, (A, B), using some arithmetic of absolute value and sign functions (we use the multivalued sign function so that $sign(0) = \{\pm 1\}$):

$$A = (|x| + |y|)sign(x)$$

$$B = \frac{y \cdot sign(x)}{|x| + |y|}$$

5.3 The Connection to the Möbius Strip

These equations define a map between the Cartesian plane *XY* to a point on the Möbius strip. We call this map the "Soft Möbius map," which is depicted in Fig. 5.7. The significance of this map is that it transforms the points on the plane into Soft numbers that are represented as points on the Möbius strip. Each colored region is mapped to a corresponding region on the Möbius strip.

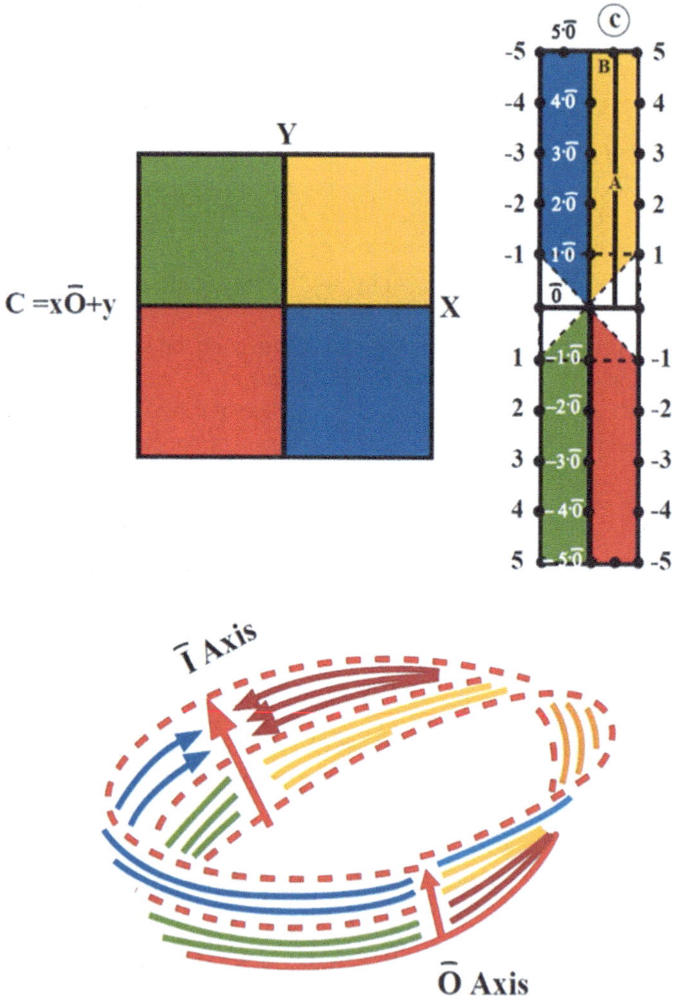

Fig. 5.7 The Soft Möbius map

Reference

Klein, M., Maimon, O.: Fundamentals of Soft logic. New Math. Nat. Comput. **17**(3), 703–737 (2021). https://doi.org/10.1142/S1793005721500356

Chapter 6
Soft Logic and Calculus

This chapter deals with the natural link between Soft logic and differential and integral calculus. When one examines the power of a Soft number, the connection to derivatives of functions is revealed. We extend this property to polynomials of Soft numbers and then to analytic functions. Subsequently, we define Soft functions, which are an expansion of real functions, and we prove the fundamental properties of these functions (Klein and Maimon 2021).

6.1 The Connection to the Derivatives Function

In this chapter, we show how Soft numbers can be used for describing and investigating the tools of calculus, the first and foremost among them being the derivative. In the previous section, we obtained the following result:

$$\left(a\bar{0}\dot{+}b\right)^n = nab^{n-1}\bar{0}\dot{+}b^n,$$

where n is any natural number. With the use of the formula for a derivative of the power function of a real variable $(x^n)' = nx^{n-1}$, the power function of a Soft variable $y\bar{0}\dot{+}x$ with a real component x can be written as $\left(y\bar{0}\dot{+}x\right)^n = nyx^{n-1}\bar{0}\dot{+}x^n = y(x^n)'\bar{0}\dot{+}x^n$.

Let us observe a polynomial function of a real variable:

$$P(x) = \sum_{k=0}^{n} a_k x^k.$$

A polynomial function of a Soft variable $y\bar{0}\dot{+}x$ is

M. Klein, O. Maimon, *Foundations of Soft Logic*,
https://doi.org/10.1007/978-3-031-58233-2_6

$$P(y\bar{0}\dot{+}x) = \sum_{k=0}^{n} a_k (y\bar{0}\dot{+}x)^k = \sum_{k=1}^{n} ka_k yx^{k-1}\bar{0} \dot{+} \sum_{k=0}^{n} a_k x^k.$$

By comparing the two polynomial functions above, we arrive at the fundamental formula connecting Soft numbers with calculus, as presented below.

Lemma 6.1 If $P(x)$ is a real polynomial function, then $P\left(y\bar{0}\dot{+}x\right) = yP'(x)\bar{0}\dot{+}P(x)$.

If $f(x)$ is an analytic function, it can be approximated by its Taylor polynomials $P_n(x)$, so that we have $\lim_{n\to\infty} P_n(x) = f(x)$, $\lim_{n\to\infty} P_n{}'(x) = f'(x)$.

On the basis of these limit conditions and the statement of Lemma 6.1, we arrive at the following definition of a function of a Soft variable that arises from the real differentiable function $f(x)$ and that we denote $\overline{f}$:

for any Soft number $y\bar{0} \dot{+} x, \overline{f}(y\bar{0} \dot{+} x) := yf'(x)\bar{0} \dot{+} f(x)$.

■

This formula may be seen as an analog of Lagrange's formula in the world of Soft numbers. It extends a real differentiable function f to a Soft function $\overline{f}$.

We can use Soft numbers to develop the product rule and the chain rule of derivatives:

$$(f \times g)(y\bar{0} \dot{+} x) =$$

$$f(y\bar{0} \dot{+} x) \times g(y\bar{0} \dot{+} x) =$$

$$(yf'(x)\bar{0} \dot{+} f(x)) \times (yg'(x)\bar{0} \dot{+} g(x)) =$$

$$(yf'(x) \times g(x) + yf(x) \times g'(x))\bar{0} \dot{+} f(x) \times g(x).$$

Hence we obtain the following product derivative rule:

$$(f \times g)' = f' \times g + f \times g'.$$

The chain rule for the derivative of a composite function is presented below:

$$(f \circ g)\left(y\bar{0} \dot{+} x\right) = f\left(g\left(y\bar{0} \dot{+} x\right)\right) =$$

$$f\left(yg'(x)\bar{0} \dot{+} g(x)\right) = yf'(g(x))g'(x)\bar{0} \dot{+} f(g(x)).$$

Hence

$$(f \circ g)' = g' \times f'(g).$$

6.2 Soft Functions

In this section, we use two different ways of extending a real function $f: \mathrm{R} \rightarrow \mathrm{R}$ to a Soft function. The first way of extension provides the function $\bar{f}$: **SN** → **SN** of a Soft variable with Soft values. In the second way, we obtain the extended function $\widehat{f}$: R → **SN** of a real variable with Soft values.

The first way of extension can be applied only to such functions $f(x)$ that are differentiable. Given a real differentiable function f, we will define its Soft extension $\bar{f}$ with the following formula (x, y are any real numbers in the domain of f):

$$\bar{f}(x\bar{0}\dot{+}y) := xf'(y)\bar{0}\dot{+}f(y).$$

6.2.1 *Fundamental Properties of Soft Functions*

In this subsection we consider the functions $\bar{f}$: **SN** → **SN**.

Lemma 6.2 For any two differentiable real functions f, g there is

(a) $\overline{f+g} = \bar{f} + \bar{g}$
(b) $\overline{f \cdot g} = \bar{f} \cdot \bar{g}$
(c) $\overline{f \circ g} = \bar{f} \circ \bar{g}$
(d) $\overline{\left[\frac{1}{f}\right]} = \frac{1}{\bar{f}}, (f \neq 0)$

Remark 6.1 A quotient with the number 1 as a numerator and a Soft number as a denominator denotes the inverse of this Soft number.

Proof 6.2

(a)

$$\begin{aligned}\overline{f+g}(x\bar{0}\dot{+}y) &= x(f+g)'(y)\bar{0}\dot{+}(f+g)(y) = (xf'(y)+xg'(y))\bar{0}\dot{+}(f(y)+g(y)) \\ &= (xf'(y)\bar{0}\dot{+}f(y)) + (xg'(y)\bar{0}\dot{+}g(y)) = \bar{f}(x\bar{0}\dot{+}y) + \bar{g}(x\bar{0}\dot{+}y).\end{aligned}$$

Therefore $\overline{f+g} = \bar{f} + \bar{g}$.

■

(b)

$$\overline{f\cdot g}(x\bar{0}\dotplus y)=x(fg)'(y)\bar{0}\dotplus(fg)(y)=(xf'(y)g(y)+xf(y)g'(y))\bar{0}\dotplus f(y)g(y)$$
$$=(xf'(y)\bar{0}\dotplus f(y))\cdot(xg'(y)\bar{0}\dotplus g(y))=\bar{f}(x\bar{0}\dotplus y)\cdot\bar{g}(x\bar{0}\dotplus y).$$

Therefore $\overline{f\cdot g}=\bar{f}\cdot\bar{g}$.

■

(c)

$$(\bar{f}\circ\bar{g})(x\bar{0}\dotplus y)=\bar{f}(xg'(y)\bar{0}\dotplus g(y))=xg'(y)f'(g(y))\bar{0}\dotplus f(g(y))$$
$$=x(f\circ g)'(y)\bar{0}\dotplus(f\circ g)(y)$$
$$=f\bar{\circ}g(x\bar{0}\dotplus y).$$

Therefore $\overline{f\circ g}=\bar{f}\circ\bar{g}$.

■

(d)

$$\overline{\left[\frac{1}{f}\right]}(x\bar{0}\dotplus y)=x\left(\frac{1}{f}\right)'(y)\bar{0}\dotplus\left(\frac{1}{f}\right)(y)$$
$$=-x\frac{f'}{f^2}(y)\bar{0}\dotplus\frac{1}{f(y)}$$
$$=\frac{1}{xf'(y)\bar{0}\dotplus f(y)}=\frac{1}{\bar{f}(x\bar{0}\dotplus y)}.$$

Thus, it is proved that $\overline{\left[\frac{1}{f}\right]}=\frac{1}{\bar{f}}$, where $f\neq 0$, and this completes the Lemma 6.2 proof.

■

Before formulating the next statement, we wish to remind the reader of the concept of an inverse function. Let f, g be real functions, each one mapping R on R. Let e be the identity function in R, mapping every real number onto itself, i.e., $e(x)=x$ for every real x. The function g is said to be an inverse function for f and denoted as $g=f^{-1}$, iff: $f\circ g=g\circ f=e$.

Let F, G be Soft functions, each one mapping **SN** on **SN**. Let I denote the identity function in **SN**, which maps every Soft number into itself, i.e., $I(x\bar{0}\dotplus y)=x\bar{0}\dotplus y$ for any real x, y. It is easy to verify that $I=\bar{e}$.

The Soft function G is said to be an inverse function for the Soft function F and denoted as $G=F^{-1}$, if $F\circ G=G\circ F=I$.

Lemma 6.3 If $g = f^{-1}$ in R, then $\overline{g} = \left(\overline{f}\right)^{-1}$ in **SN**.

Proof 6.3 If $g = f^{-1}$, then $f \circ g = g \circ f = e$, and consequently $\overline{f \circ g} = \overline{g \circ f} = \overline{e} = I$. By Lemma 6.2 c, $\overline{f \circ g} = \overline{f} \circ \overline{g}$. Combining this equation with a previous one, we obtain $\overline{f} \circ \overline{g} = \overline{g} \circ \overline{f} = I$, which means that $\overline{g} = \left(\overline{f}\right)^{-1}$.

■

Lemma 6.4 For any two Soft numbers $\alpha = a\overline{0} \dotplus b$, $\beta = c\overline{0} \dotplus d$, there exists $\overline{e^{\alpha+\beta}} = \overline{e^{\alpha}}\,\overline{e^{\beta}}$.

Proof 6.4 According to the rule of an extension of a real function to a Soft one, we have on one side:

$$\overline{e^{\alpha+\beta}} = \overline{e^{\left(a\overline{0} \dotplus b\right) + \left(c\overline{0} \dotplus d\right)}} = \overline{e^{(a+c)\overline{0} \dotplus (b+d)}} = (a+c)e^{b+d}\overline{0} \dotplus e^{b+d},$$

and on the other side:

$$\begin{aligned}\overline{e^{\alpha}}\ \overline{e^{\beta}} &= \overline{e^{(a\overline{0}\, \dotplus\, b)}}\ \ \overline{e^{(c\overline{0}\, \dotplus\, d)}} = \left(ae^{b}\overline{0} \dotplus e^{b}\right)\left(ce^{d}\overline{0} \dotplus e^{d}\right) \\ &= (a+c)e^{b+d}\overline{0} \dotplus e^{b+d}.\end{aligned}$$

By comparing the two results above, we conclude $\overline{e^{\alpha+\beta}} = \overline{e^{\alpha}}\ \overline{e^{\beta}}$.

■

Reference

Klein, M., Maimon, O.: Fundamentals of Soft logic. New Math. Nat. Comput. **17**(3), 703–737 (2021). https://doi.org/10.1142/S1793005721500356

Chapter 7
Soft Curves

In this chapter we define Soft curves, which describe real functions, by using the Soft coordinate system. We present several examples of Soft curves and then prove the fundamental equation of Soft curves and the symmetry of curves described by a function and its inverse. Finally, we prove the link between Soft curves and the calculation of the rectangular areas created by them (Klein and Maimon 2021).

7.1 Definition of Soft Curves

In this chapter, we use the second way of extending a real function f to a Soft function of a real variable $\widehat{f}: \mathrm{R} \to \mathrm{SN}$, performed by the formula $\widehat{f}(x) = x\overline{0} \dotplus f(x)$, which does not require differentiability of f. Continuing on this path, we construct a graphic presentation of f in the Soft numbers coordinate system, defined in the SNS (the Soft number strip).

Any Soft number $x\overline{0} \dotplus y$ with $xy > 0$ is represented on the SNS as a point with a height A and a width B:

$$A = x + y,$$

$$B = \frac{y}{x + y}.$$

Applying this assertion to a Soft function $\widehat{f}(x)$, we come to the conclusion that any Soft function $\widehat{f}(x) = x\overline{0} \dotplus f(x)$ of a real variable x that satisfies the condition $xf(x) > 0$ can be visually presented on the SNS as a curve described by the following equations:

M. Klein, O. Maimon, *Foundations of Soft Logic*,
https://doi.org/10.1007/978-3-031-58233-2_7

$$A = x + y = x + f(x),$$

$$B = \frac{y}{x+y} = \frac{f(x)}{x+f(x)} \quad x \neq -y \ , x \neq -f(x)$$

(Note that $y = f(x)$ is the equation of an original real function $f(x)$ generating a Soft function $\widehat{f}(x)$).

As $\widehat{f}(x)$ is a Soft extension of $f(x)$, the presentation of a Soft function $\widehat{f}(x)$ on the SNS will be called by us as a presentation of a real function $y = f(x)$ on the SNS, or, briefly, a Soft presentation of $y = f(x)$. For the construction of such a presentation, we define an orthogonal (A, B) coordinate system on the SNS, where A is a height, B is a width of a point on the SNS, the A-axis coincides with the zero line, and the B-axis is a horizontal segment starting at the absolute zero and continuing to the right real line.

Let us present two examples of constructing a Soft graphic presentation of real functions in the (A, B) coordinate system on the SNS.

Example 7.1 $f(x) = kx$, where $k>0$ and x satisfy the condition: $xf(x) > 0 \Leftrightarrow kx^2 > 0 \Leftrightarrow x \neq 0$. For any $x \neq 0$ we have

$$A = x + kx = (k+1)x,$$

$$B = \frac{kx}{x+kx} = \frac{k}{k+1}, k \neq -1.$$

Because B is constant, we obtain two straight lines on the SNS that are parallel to the zero axis at a distance $B = \frac{k}{k+1}, k \neq -1$ from it.

In Fig. 7.1, we can see the presentation $f(x) = kx$, $k > 0$ in the Soft coordinate system.

Example 7.2 $f(x) = kx + d$, where $k > 0, d > 0, x > 0$ or $x < -\frac{d}{k}$.

Under these conditions, we have

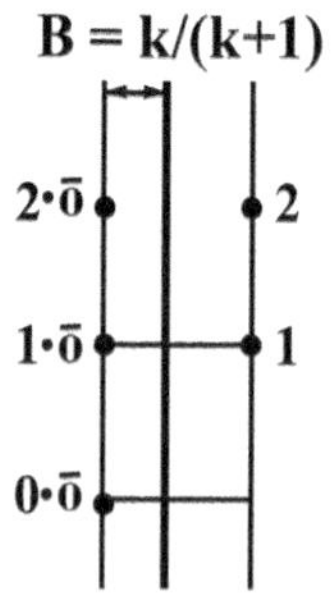

Fig. 7.1 The Soft curve $y = f(x) = kx, k > 0$

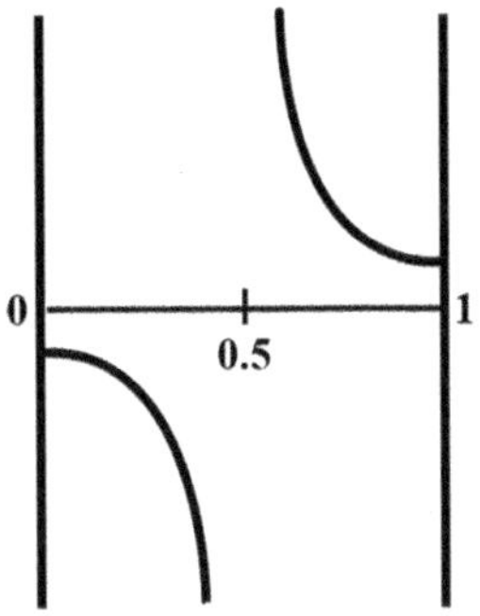

Fig. 7.2 The Soft curve $y = kx + d$

$$A = (k + 1)x + d,$$

$$B = \frac{kx + d}{(k + 1)x + d}, (k + 1)x \neq -d.$$

From these two equations, A can be expressed in terms of B by the following calculation: $x = \frac{A-d}{k+1} \Rightarrow AB = \frac{k(A-d)}{k+1} + d \Rightarrow A = \frac{d}{(k+1)B-k}$.

The final equation leads to the conclusion that the graphic presentation of the function $f(x) = kx + d$ on the SNS under the given restrictions is a hyperbolic curve with two branches and a vertical asymptote $B = \frac{k}{k+1}$.

In the case $k = d = 1$, the result looks as shown in Fig. 7.2.

Examples 7.1 and 7.2 show that the development of the equation of the Soft graphic presentation of $f(x)$ takes time and effort, even in the case of simple functions $f(x)$. This problem is reduced by the following Theorem 7.1.

7.2 The Fundamental Equation for Soft Curves

Theorem 7.1 Let $f(x)$ be any real function satisfying the condition $xf(x) > 0$ in some real domain D. Then in this domain, the function $f(x)$ can be presented on the SNS with a curve that has in the coordinates A, B the following equation:

$AB = f(A - AB)$. If the function f has an inverse function f^{-1}, then the equation above is equivalent to the equation $A - AB = f^{-1}(AB)$.

Proof 7.1 By definition,

$$\begin{aligned} A &= x + f(x), \\ B &= \frac{f(x)}{A}. \end{aligned}$$

hence

$$AB = f(x).$$

These equations lead to

$$\begin{aligned} x &= A - AB, \\ f(x) &= f(A - AB). \end{aligned}$$

Substituting this expression for $f(x)$ into the equation $AB = f(x)$, we obtain $AB = f(A - AB)$.

Thus we obtain the first equation of the theorem. If the function f has an inverse function, then by applying it to both sides of the above equation, we obtain the second equation of the theorem, which is equivalent to the first one.

■

As the first equation in Theorem 7.1 is valid for any real function f, we will call it "a general (A, B) equation of $f(x)$," while the second equation will be called "an inverse (A, B) equation of $f(x)$." Let us demonstrate the use of Theorem 7.1 for the functions in the Examples 7.1 and 7.2 above.

The general (A, B) equation of $f(x) = kx$ $(k > 0)$ is

$$AB = k(A - AB) \Rightarrow B = \frac{k}{k+1}, k \neq -1.$$

The general (A, B) equation of $f(x) = kx + d$ $(k > 0, d > 0)$ is

$$AB = k(A - AB) + d \Rightarrow A = \frac{d}{(k+1)B - k}.$$

We see that with the help of Theorem 7.1, the final equations in Examples 7.1 and 7.2 are obtained faster and easier than before, without it. Therefore, in the next examples we will use Theorem 7.1 as a permanent tool for obtaining (A, B) equations of given real functions. But before this, we wish to note several important properties of the equations stated in Theorem 7.1

Remark 7.1 The equations in Theorem 7.1 are implicit. If A can be expressed in terms of B, then equivalent explicit equations are obtained. Either an implicit or an explicit form of an equation with two variables A, B allows the construction of corresponding curve on the SNS.

Remark 7.2 The general (A, B) equation of $f(x)$ can be obtained from the original equation $y = f(x)$ by the substitutions $x = A - AB$, $y = AB$. These are the formulas by which a Soft number $x\bar{0}\dot{+}y$ is assigned to any point C with a nonzero height A and (since $xf(x) > 0$) a positive width B on the SNS.

Now we can present and discuss examples of constructing Soft presentations of real functions by exploring Theorem 7.1. At this stage, two preparatory remarks are needed.

(1) Any function $f(x)$ in the examples below is considered under the condition $xf(x) > 0$, which guarantees the existence of the parameters A, B. Under this condition, the graphic presentation of $f(x)$ on the SNS is located in the inner space between the lines $B = 0$, $B = 1$, and $A = 0$. Nevertheless, some curves in the following examples contain points on these lines. Such points can be seen in quite a regular way as limit positions of the inner presentation, when one of the variables x, $y = f(x)$ or both of them tend to 0.

(2) Since we restrict the domain of all functions to ensure that they satisfy the condition $xf(x) > 0$, the Soft graphic presentations here will be purely to the right of the zero axis. One can prove an analogous theorem to Theorem 7.1 for functions for which $xf(x) < 0$, and the graphic presentation given thereby is to the left of the zero axis. We will not investigate this here.

Example 7.3 $y = x^i$ where $i = 2, 3, 4, \ldots$.

In this case $f(x) = x^i$, and the admissible values of x are $xf(x) > 0 \Leftrightarrow x^{i+1} > 0 \Rightarrow x \neq 0$ for an odd i, and then $x > 0$ for an even i.

By Theorem 7.1, the general (A, B) equation for the given function is $AB = (A - AB)^i$. From this equation we can, by obvious algebraic transformations, obtain the explicit expression $A = \sqrt[i-1]{\frac{B}{(1-B)^i}}$.

In Fig. 7.3, we see Soft graphs created by this equation on the SNS for $i = 2, 3, \ldots, 10$. For $i = 3, 5, 7, 9$, the degree $i-1$ of the root in the formula for A is even, and therefore each one of the four graphs corresponding to these i's has two branches: a positive one and a negative one, symmetric about the segment $A = 0$.

Example 7.4 $y = \log_i x$, $i = 2, 3, 4, \ldots$.

In this case, $f(x) = \log_i x$, and the condition $xf(x) > 0 \Leftrightarrow x\log_i x > 0$ is satisfied if $x > 1$. For such values of x, by Theorem 7.1, the general (A, B) equation of $f(x)$ is $AB = \log_i(A - AB)$. This implicit equation, for $i = 2, 3, \ldots, 10$, defines nine curves on the SNS as shown in Fig. 7.4.

Example 7.5 $y = \tan x$.

In this case, the admissible values of x are

$$xf(x) > 0 \Rightarrow x \tan x > 0 \Leftrightarrow k\pi < x < k\pi + \frac{\pi}{2},$$

$$-k\pi - \frac{\pi}{2} < x < -k\pi \quad (k = 0, 1, 2, 3, \ldots).$$

For any such x, the values of a height A and a width B do exist and are defined by the following formulas:

$$A = x + \tan x,$$

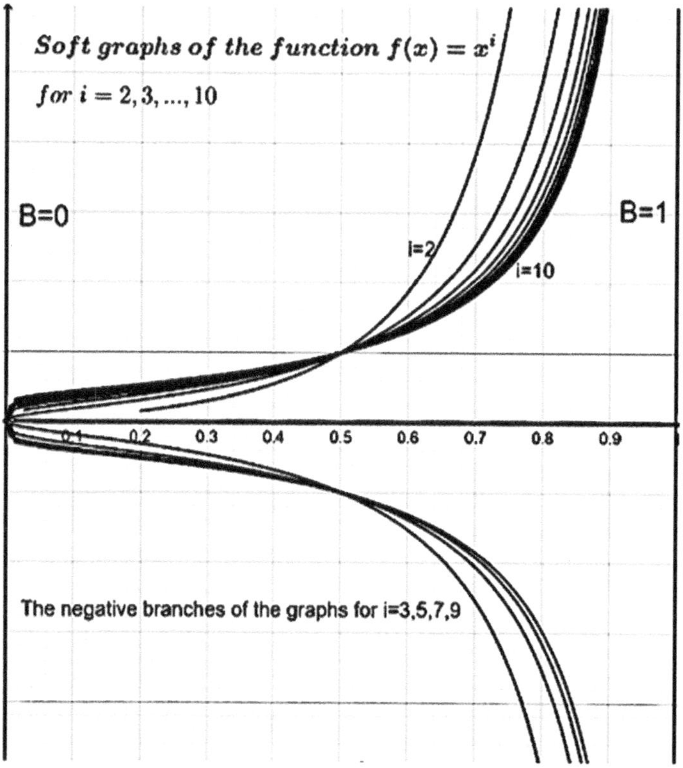

Fig. 7.3 The Soft curve $y = x^i$

$$B = \frac{\tan x}{x + \tan x}, x \neq -\tan x.$$

Conceiving x as a parameter and the two equations above as a parametric description of the graphic presentation of tan(x) in the (A, B) coordinate system on the SNS, we can investigate the behavior of this presentation in any of the intervals indicated above. The investigation results are as follows:

In the interval $0 < x < \frac{\pi}{2}$, the graph ascends from the point $A = 0$, $B = 1/2$ to the line $B = 1$ as an asymptote. In every interval $k\pi < x < k\pi + \frac{\pi}{2}$ $(k = 1, 2, 3, \ldots)$, the graph ascends from the point $A = k\pi$, $B = 0$ to the line $B = 1$ as an asymptote. In every interval $-k\pi - \frac{\pi}{2} < x < -k\pi$, the graph is symmetric about the segment $A = 0$, to the graph in the interval $k\pi < x < k\pi + \frac{\pi}{2}$ $(k = 0, 1, 2, 3, \ldots)$.

(The interested reader is invited to verify these assertions). As the number of intervals of admissible values of x for *tan*(x) is infinite, there is an infinite number of curves in the Soft graphic presentation of $y = \tan x$ on the SNS. In Fig. 7.5, only a finite part of the whole picture on the SNS is shown.

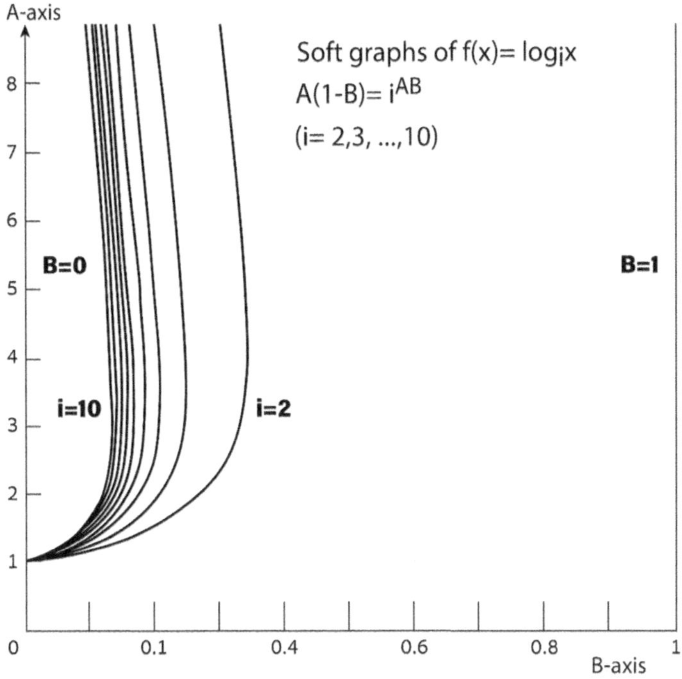

Fig. 7.4 The Soft curve $y = \log_i x$

The separate branches of the Soft graph $y =$ tan x are created with the help of the inverse (A, B) equations:

$$A - AB = \arctan(AB) + m\pi \quad (m = 0, \pm 1, \pm 2, \ldots).$$

■

Example 7.6 $f(x) = e^x$

The general (A, B) equation of this function in the domain $x > 0$ is $AB = e^{A - AB}$.

The Soft graph of this function on the SNS is presented in Fig. 7.10.

Example 7.7 $f(x) = \sqrt{x}$.

The general (A, B) equation of this function in the domain $x>0$ is $AB = \sqrt{A - AB}$, and its inverse (A, B) equation is $A - AB = (AB)^2$.

The Soft graph of $f(x) = \sqrt{x}$ is presented in Fig. 7.6.

Example 7.8 $f(x) = \frac{1}{x}$.

The general (A, B) equation of this function in the domain $R\backslash\{0\}$ is $AB = \frac{1}{A - AB}$, which is equivalent to $A^2B - A^2B^2 = 1$.

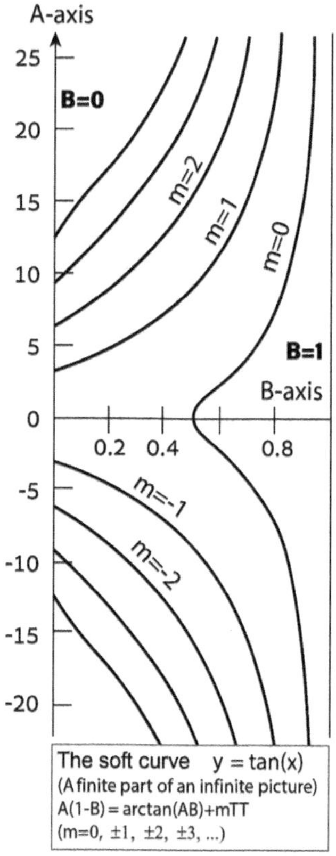

Fig. 7.5 The Soft curve $y = \tan x$

The corresponding Soft graphic presentation on the SNS is shown in Fig. 7.7.

Example 7.9 $f(x) = \frac{x}{1+x^2}$.

It is an odd continuous function on R, satisfying the condition $xf(x) > 0$ for all real nonzero x. In this domain, it has no inverse function and consequently no inverse (A, B) equation. The general (A, B) equation of this function is

$$AB = \frac{A - AB}{1 + (A - AB)^2}.$$

The corresponding Soft presentation on the SNS is shown in Fig. 7.8.

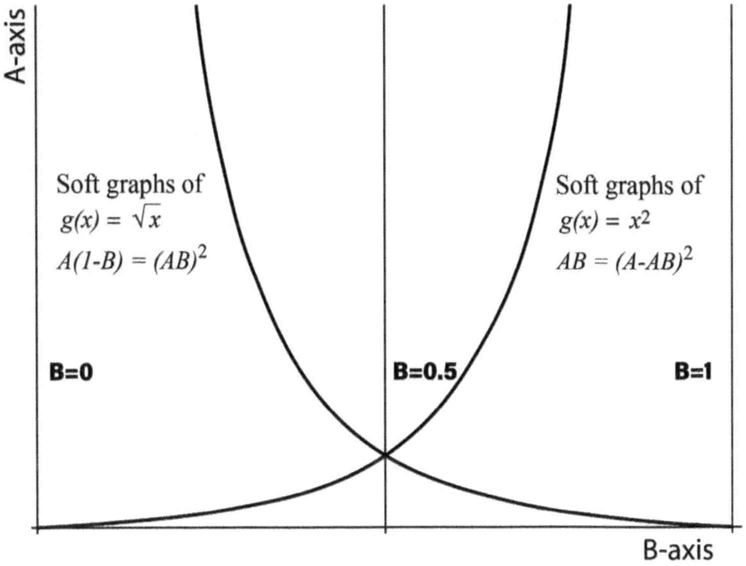

Fig. 7.6 The Soft curve $f(x) = \sqrt{x}$

Fig. 7.7 The Soft curve $f(x) = \frac{1}{x}$

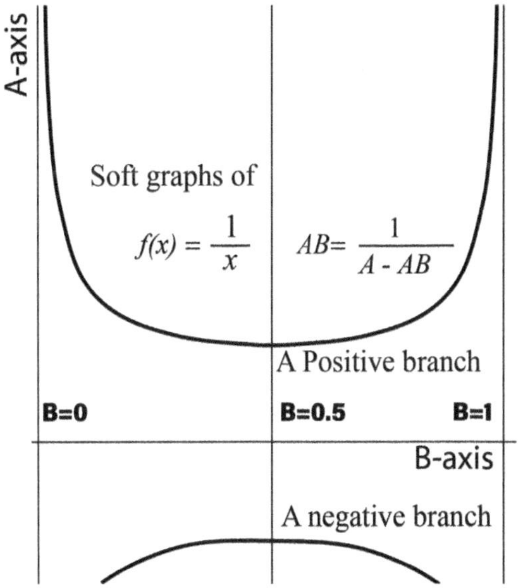

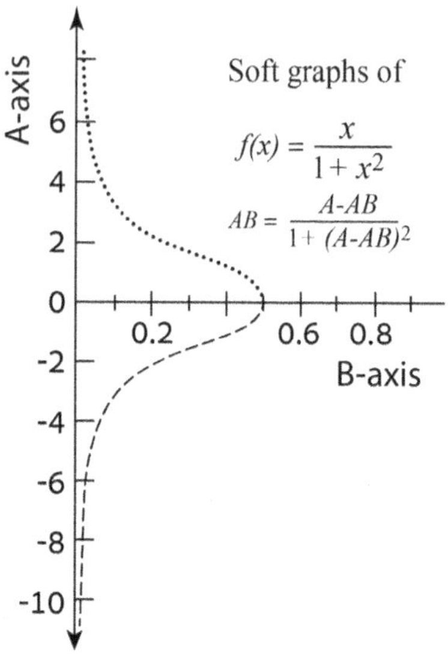

Fig. 7.8 The Soft curve $f(x) = \frac{x}{1+x^2}$

7.3 Presentation of the Symmetry of Two Mutually Inverse Functions

In this subsection, we will discuss the graphic presentation on the SNS of two real functions that are inverse to each other, e.g., $f(x) = e^x$, $g(x) = \ln x$. In the Cartesian coordinate system (x, y), their graphs are symmetric about the line $y = x$ as shown in Fig. 7.9.

Figure 7.10 shows the graphic images of the same two functions in the Soft numbers coordinate system.

We can see that there is a mirror symmetry about the vertical line $B = 1/2$ in the Soft graphic presentations of these two functions. It may be verified that this situation is valid for any two mutually inverse real functions.

Theorem 7.2 If f, g are two real functions, inverse to each other, then the right parts of their Soft presentations in the same SNS are symmetric about the vertical line $B = 1/2$.

Proof 7.2 By Theorem 7.1, two equivalent (A, B) equations of the Soft graph of $f(x)$ on the SNS are

$$AB = f(A(1 - B)),$$

$$A(1 - B) = g(AB),$$

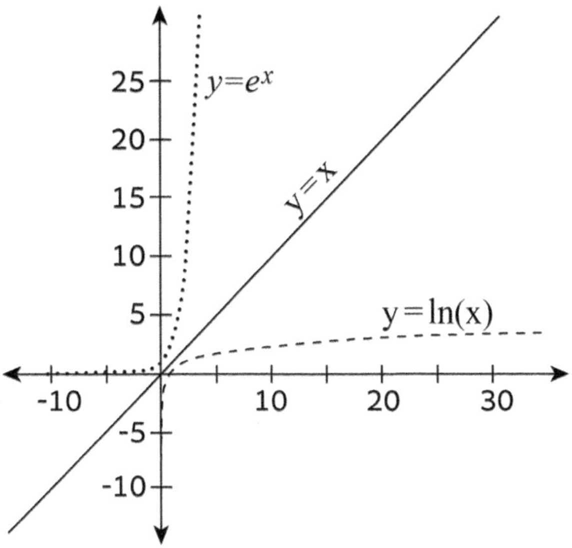

Fig. 7.9 $f(x) = \ln x$, $g(x) = e^x$

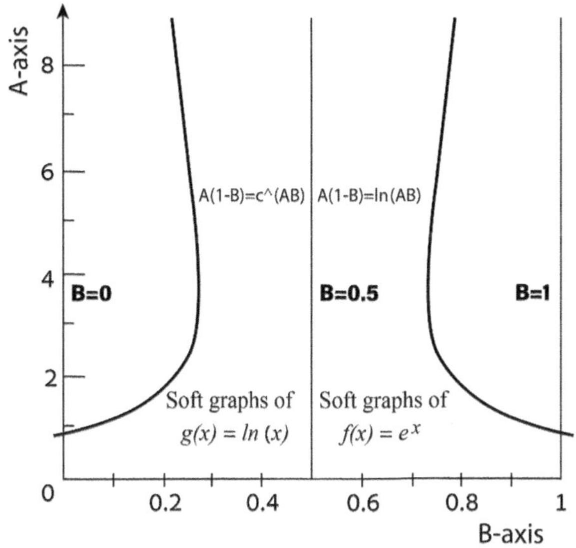

Fig. 7.10 $\ln x$, $g(x) = e^x$ in the Soft coordinate system

and two equivalent (A, B) equations of the Soft graph of $g(x)$ are

$$AB = g(A(1 - B)),$$

$$A(1 - B) = f(AB).$$

It can be easily seen that the mapping $(A, B) \to (A, 1 - B)$ transforms one pair of the above equations into the other pair. As such a mapping transfers any point in the (A, B) coordinate system into a point symmetric to it about the line $B = 1/2$, the conclusion is that the curves representing two mutually inverse functions f, g in the right part of the SNS are symmetric about the vertical line $B = 1/2$.

■

To resume and exemplify this statement, let us return to the functions $f(x) = e^x$ and $g(x) = \ln x$ that are inverse to each other. The curves representing them on the SNS are symmetric about the line $B = 1/2$ (Fig. 7.10). The same situation exists for the functions $f(x) = x^2, g(x) = \sqrt{x}$ in the domain $x > 0$ (Fig. 7.6).

The case of the function $f(x) = \frac{1}{x}$ is special, because this function and its inverse function are identical:

$$f^{-1}(x) = f(x) = \frac{1}{x}.$$

Therefore, the Soft graphic presentation of $f(x) = \frac{1}{x}$ in the SNS is itself symmetric about the line $B = 1/2$ (Fig. 7.7).

7.4 Soft Curves and Areas

A general (A, B) equation of a real function f on the SNS can be seen as a very simple equation between some areas in the SNS. Let $C\,(A, B)$ be any fixed point in the (A, B) coordinate system on the right part of the SNS, with $A > 0$ or $A < 0$ and $0 < B < 1$. Two segments containing the point C can be drawn: a horizontal segment going through the point C and connecting the zero line with the right real line and a vertical segment going from the point C to the segment $A = 0$.

With this drawing, two rectangles are created. The left rectangle has a vertical side with a length A on the zero line, and the length of its horizontal side is B. The right rectangle has a vertical side with a length A on the right real line, and the length of its horizontal side is $1 - B$. Let S_0, S_1 denote the areas of these rectangles, as shown in Fig. 7.11.

The above construction enables visualizing the general (A, B) equation of a real function f, as stated in the following Lemma.

Theorem 7.3 A general (A, B) equation on the SNS of any real function f can be written in the form $S_0 = f(S_1)$, where S_0, S_1 are the areas shown in Fig. 7.11 (if $A < 0$, the areas are denoted with a minus sign).

***Proof** 7.3* The definition of S_0, S_1 requires that $S_0 = AB$, $S_1 = A(1 - B)$, and therefore the equation $S_0 = f(S_1)$ is just another (visual) form of the general (A, B) equation of f, stated in Theorem 7.1.

■

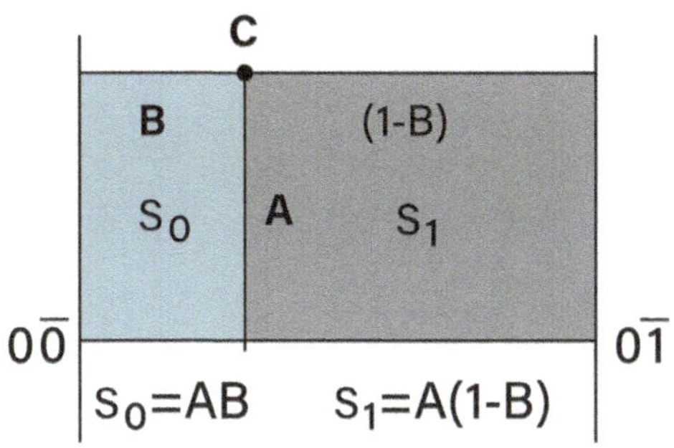

Fig. 7.11 Rectangles of the Soft coordinate system

It is worth noting here that the areas indicated in Fig. 7.11 also visualize the real factors of the Soft number $x\bar{0}\dot{+}y$, which is represented by the point on the SNS with a height A and a width B, namely, $x = S_1$, $y = S_0$.

Reference

Klein, M., Maimon, O.: Fundamentals of Soft logic. New Math. Nat. Comput. **17**(3), 703–737 (2021). https://doi.org/10.1142/S1793005721500356

Chapter 8
The Dynamics of Soft Logic

In this chapter, we define the dynamics of Soft functions, meaning what occurs when we repeatedly apply a Soft function on an initial number. We describe the mathematical properties of the sequence of numbers created in this process, and we show the connection that is created between the dynamics of Soft functions and high-order derivatives. We present several examples of the dynamics of specific functions. In conclusion, we describe geometrically the dynamics of Soft functions, as inspired by fractal and chaos theory (Klein and Maimon 2020).

8.1 Dynamics with Soft Numbers

Given a differentiable real function f, defined on **R**, we can extend it to the set of all Soft numbers **SN** and then apply it again and again, starting with any chosen and fixed Soft number C_0 and continuing with its successors. In this way, we generate an infinite sequence of Soft numbers. We wish to investigate the behavior of the infinite sequence $C_0, C_1, \ldots, C_n, \ldots$, where

$$\begin{aligned} C_0 &= X_0\bar{0}\dot{+}Y_0, \\ C_1 &= f(C_0) = f\left(X_0\bar{0}\dot{+}Y_0\right) \\ &= X_0 f'(Y_0)\bar{0}\dot{+}f(Y_0), \\ C_2 &= f(C_1) \\ &= X_0 f'(Y_0) f'(f(Y_0))\bar{0}\dot{+}f(f(Y_0)) \end{aligned}$$

and so on.

Definition 8.1 Let a real function $f(x)$, defined on R, be given. The composite functions obtained from it by the recursion $f_0(x) = x$, $f_1(x) = f(x)$, $f_k(x) = f(f_{k-1}(x))$ $(k = 2, 3, \ldots)$ are called iterations of $f(x)$.

M. Klein, O. Maimon, *Foundations of Soft Logic*,
https://doi.org/10.1007/978-3-031-58233-2_8

Lemma 8.1 Let $f(x)$ be any differentiable real function on **R**, and let $f(C)$ denote its Soft extension on **SN**. If

$$C_0 = X_0\bar{0} \dotplus Y_0,$$
$$C_k = f(C_{k-1}), \qquad k = 1, 2, 3, \ldots,$$

then

$$C_k = X_0 f_k{}'(Y_0)\bar{0} \dotplus f_k(Y_0), \;\; k = 1, 2, 3, \ldots.$$

Proof 8.1 We first prove by induction that for every natural k we have $C_k = f_k(C_0)$. Indeed, $f_0(C_0) = C_0$ and $C_k = f(C_{k-1}) = f(f_{k-1}(C_0)) = f_k(C_0)$. Now, from the definition of the Soft extension of a real function, it follows that

$$C_k = f_k(C_0) = X_0 f_k{}'(Y_0)\bar{0} \dotplus f_k(Y_0)$$

For every natural k.

■

Lemma 8.2 Let functions $f(x)$, $f(C)$ be as defined in Lemma 8.1.
If the sequence of real numbers

$$D_k = \frac{f_k{}'(Y_0)}{f_k(Y_0)} \;\; (k = 1, 2, 3, \ldots)$$

tends to 0, then the sequence of points on the SNS, which are the right parts of the images of the Soft numbers C_k defined in Lemma 8.1, tends to the 1 axis. If the sequence D_k tends to infinity and $X_0 > 0$, then the sequence of these points on the SNS tends to the zero axis.

Proof 8.2 If the "convex" method of the presentation of Soft numbers as points on the SNS is used, then the distance of the points presenting the Soft numbers C_k on the SNS to the zero axis is given by the equation:

$$B_k = \frac{Y_k}{X_k + Y_k} = \frac{1}{\frac{X_k}{Y_k} + 1} =$$

$$= \frac{1}{\frac{X_0 f_k{}'(Y_0)}{f_k(Y_0)} + 1} = \frac{1}{X_0 D_k + 1}, X_k \neq Y_k.$$

The lemma's statements are immediately obtained from the final expression for B_k in the above development.

■

Remark 8.1 In the previous considerations, it was supposed that a real function *f* is defined over the whole set **R** of real numbers. The obtained results may be generalized for a real function defined in any non-empty set *E* of real numbers, if the function values also belong to *E*.

Remark 8.2 In the following examples, the "convex" presentation of Soft numbers on the SNS is used, when the condition of its existence is fulfilled. In such a case, a sequence of Soft numbers is conceived as a sequence of points in the right part of the SNS. The location of a point on the SNS is stated by its height *A* and width *B*. Therefore, the investigation of the behavior of point sequences on the SNS is reduced to finding and examining the sequences of real numbers A_k, B_k that state the point's location on the SNS.

The functions *f* in the Examples 8.1, 8.2, 8.3, 8.4 and 8.5 are defined directly as operations in **SN**, while in Examples 8.6, 8.7 and 8.8, the functions are defined as extensions of the corresponding real functions.

8.2 Examples of the Dynamic with Soft Numbers

Example 8.1 Let a be any given real nonzero number. Let us define $f(C) = aC$ for any Soft number C, such that

$$C_0 = X_0\bar{0} \dotplus Y_0,$$
$$C_1 = f(C_0) = aX_0\bar{0} \dotplus aY_0,$$
$$C_k = f(C_{k-1}) = a^k X_0\bar{0} \dotplus a^k Y.$$

Consequently,

$$A_k = a^k X_0 + a^k Y_0 = a^k(X_0 + Y_0),$$
$$B_k = \frac{a^k Y_0}{a^k(X_0 + Y_0)} = \frac{Y_0}{X_0 + Y_0} = \frac{1}{1 + \frac{X_0}{Y_0}}$$
$$X_0 \neq -Y_0$$

By the last equation, the presentation of the Soft numbers C_k as points on the SNS exists if $X_0Y_0 > 0$, or $X_0 = 0$, $Y_0 \neq 0$. In these cases, all the points are of equal distance to the zero axis, i.e., they are positioned on one vertical line. When $|a| > 1$, the points tend to infinity, and when $|a| < 1$, the points tend to the limit point $A = 0, B = \frac{Y_0}{X_0 + Y_0}$.

Example 8.2 For any given real $b \neq 0$ and for any Soft number C, let us define the following:

$$
\begin{aligned}
f(C) &= C + b, \\
C_0 &= X_0\bar{0} \dotplus Y_0, \\
C_1 &= X_0\bar{0} \dotplus Y_0 + b, \\
C_k &= X_0\bar{0} \dotplus Y_0 + kb.
\end{aligned}
$$

Consequently,

$$
\begin{aligned}
A_k &= X_0 + Y_0 + kb, \\
B_k &= \frac{Y_0 + kb}{X_0 + Y_0 + kb}.
\end{aligned}
$$

Suppose $Y_0 > 0, b > 0$. If $X_0 > 0$, the sequence of points presenting C_k on the SNS exists and tends to infinity and to the 1 axis. If $X_0 = 0$, then all the points are located on the 1 axis, tending to infinity as well. If $X_0 < 0$, the sequence of points on the SNS does not exist.

Example 8.3 Let a, b be any given real nonzero numbers and C any Soft number. We define the following:

$$
\begin{aligned}
f(C) &= aC + b, \\
C_0 &= X_0\bar{0} \dotplus Y_0, \\
C_1 &= aX_0\bar{0} \dotplus aY_0 + b, \\
C_2 &= a^2X_0\bar{0} \dotplus a^2Y_0 + ab + b, \\
C_3 &= a^3X_0\bar{0} \dotplus a^3Y_0 + a^2b + ab + b, \\
C_k &= a^kX_0\bar{0} \dotplus a^kY_0 + a^{k-1}b + \cdots + ab + b.
\end{aligned}
$$

If $a \neq 1$,then $C_k = a^kX_0\bar{0} \dotplus a^kY_0 + b\left(\frac{a^k-1}{a-1}\right)$.

Hence,

$$
\begin{aligned}
A_k &= a^kX_0 + a^kY_0 + b(\frac{a^k-1}{a-1}), \\
B_k &= \frac{a^kY_0 + b(\frac{a^k-1}{a-1})}{a^kX_0 + a^kY_0 + b(\frac{a^k-1}{a-1})}.
\end{aligned}
$$

We suppose that X_0, Y_0, a, b are real numbers such that the condition $0 \le B_k \le 1$ is satisfied, and the point presentation of the numbers C_k on the SNS exists. In particular, if $a > 1$, and the values of b, X_0, Y_0 are positive, such a presentation does exist, and its points tend to infinity and to the vertical line $B = \frac{Y_0+\frac{b}{a-1}}{X_0+Y_0+\frac{b}{a-1}}$.

For $0 < a < 1$ and the same condition relative to b, X_0, Y_0, the points tend to the limit point $A = \frac{b}{1-a}, B = 1$.

Example 8.4 For any Soft number C, let us define the following:

$$f(C) = C^2 = \left(X\bar{0} \dot{+} Y\right)^2 = 2XY\bar{0} \dot{+} Y^2.$$

If $C_0 = X_0\bar{0} \dot{+} Y_0$, the recursive sequence created by $f(C)$ is

$$C_1 = 2X_0Y_0\bar{0} \dot{+} Y_0^2,$$
$$C_2 = 4X_0Y_0^3\bar{0} \dot{+} Y_0^4,$$
$$C_k = 2^kX_0Y_0^{2^k-1}\bar{0} \dot{+} Y_0^{2^k}.$$

Therefore,

$$A_k = 2^kX_0Y_0^{2^k-1} + Y_0^{2^k},$$
$$B_k = \frac{Y_0^{2^k}}{2^kX_0Y_0^{2^k-1} + Y_0^{2^k}} = \frac{1}{2^k\dfrac{X_0}{Y_0} + 1}.$$

Conclusions If $X_0Y_0 > 0$,then when k tends to infinity, the distance of the points to the zero axis tends to zero. If $X_0 = 0,\ Y_0 \neq 0$,then $B_k = 1$,which means that all the points are on the 1 line. In both the cases above, if $|Y_0| > 1$, the points tend to infinity; if $|Y_0| < 1$, the point sequence tends to the limit point $A = 0, B = 1$ if $X_0 = 0$ and $A = B = 0$ otherwise.

The case $X_0 = 0, |Y_0| = 1$ leads to the point $A = 1, B = 1$, which presents a fixed point in the dynamical system of the Soft function $f(C) = C^2$.

Example 8.5 We generalize the function in Example 8.4 to any natural power n:

$$f(C) = C^n = \left(X\bar{0} \dot{+} Y\right)^n = nXY^{n-1}\bar{0} \dot{+} Y^n.$$

In this case the development, similar to the one in the previous example, gives the following:

$$A_k = n^kX_0Y_0^{n^k-1} + Y_0^{n^k},$$
$$B_k = \frac{Y_0^{n^k}}{n^kX_0Y_0^{n^k-1} + Y_0^{n^k}} = \frac{1}{n^k\dfrac{X_0}{Y_0} + 1}.$$

Conclusions For any natural power $n > 1$, and $X_0Y_0 > 0$, the sequence of points presenting the Soft numbers C_k on the SNS tends to the zero axis. If $X_0 > 0$, $Y_0 > 1$, the sequence simultaneously tends to infinity, and if $X_0 = 0$, $0 < Y_0 < 1$, the sequence is located on the 1 line and has a limit point $A = 0$, $B = 1$.

Example 8.6 A Soft power function of a Soft variable C, for any positive real power a, is defined as an extension of a real power function $f(x) = x^a$, which implies $f(C) = C^a = \left(X\bar{0}\dot{+}Y\right)^a = aXY^{a-1}\bar{0}\dot{+}Y^a$.

It is to be noted that the real function $f(x) = x^a$ with a real power a is defined only for nonnegative values of x. As its own values are also nonnegative, there exists an infinite sequence of its iterations (see Remark 8.1 after the Lemma 8.2 proof). These iterations are $f_k(x) = x^{a^k}$, k = 1, 2, 3, ...

The points on the SNS that present the Soft numbers C_k,which are created recursively by the function $f(C)$ for k = 1, 2, 3, ... starting at $C_0 = X_0\bar{0}\dot{+}Y_0$ have the following heights and widths:

$$A_k = X_0 f_k{}'(Y_0) + f_k(Y_0) = a^k X_0 Y_0^{a^k - 1} + Y_0^{a^k},$$

$$B_k = \frac{f_k(Y_0)}{X_0 f_k{}'(Y_0) + f_k(Y_0)} = \frac{Y_0^{a^k}}{a^k X_0 Y_0^{a^k - 1} + Y_0^{a^k}} = \frac{1}{a^k \dfrac{X_0}{Y_0} + 1}.$$

(Remember that in the above equations, the value of Y_0 must be nonnegative, as prescribed by the definition of the real power function).

Lemma 8.3 Let X_0, Y_0, a be any given positive numbers.

(1) If $a > 1$, then the sequence of points on the SNS, presenting the Soft numbers C_k created as defined above, tends to the zero axis.
(2) If $0 < a < 1$, then the point sequence tends to the 1 axis.

Proof 8.3 The statements of Lemma 8.3 are immediately obtained from the final expression for B_k in the development up to its formulation.

■

Example 8.7 Here we define a Soft exponent $f(C) = e^C$, where $C = X\bar{0}\dot{+}Y$, as an extension of a real function $f(x) = e^x$,by the formula $f(C) = e^c = Xe^Y\bar{0}\dot{+}e^Y$.

Given an initial Soft number C_0, this function recursively creates an infinite sequence of Soft numbers, as shown below:

$$C_0 = X_0\bar{0} \dot{+} Y_0,$$
$$C_1 = X_0 e^{Y_0}\bar{0} \dot{+} e^{Y_0},$$
$$C_2 = X_0 e^{Y_0} e^{e^{Y_0}}\bar{0} \dot{+} e^{e^{Y_0}},$$
$$C_3 = X_0 e^{Y_0} e^{e^{Y_0}} e^{e^{e^{Y_0}}}\bar{0} \dot{+} e^{e^{e^{Y_0}}},$$

and so on. The continuation of the iterative process leads, by Lemma 8.1, to the following sequence of Soft numbers:

$$C_k = X_k\bar{0} \dotplus Y_k = X_0 f_k{}'(Y_0)\bar{0} \dotplus f_k(Y_0), k = 1, 2, 3, \ldots,$$

where $f_k(x)$ is the k-th iteration of the function $f(x) = f_1(x) = e^x$. This iteration may be written in the form $f_k(x) = e \uparrow e \uparrow \ldots \uparrow e \uparrow x$, with k letters "e" in the record.

From this presentation, by the chain rule, we obtain $f_k'(x) = f_k(x) f_{k-1}(x) f_{k-2}(x) \ldots f_1(x)$.

We can now see that if k tends to infinity, both $f_k'(Y_0)$ and $f_k(Y_0)$, and their quotient $D_k = \frac{f_k'(Y_0)}{f_k(Y_0)} = f_{k-1}(Y_0) f_{k-2}(Y_0) \ldots f_1(Y_0)$ tend to infinity for any Y_0. Let us now examine three cases for X_0:

(1) $X_0 > 0$. In this case, by Lemma 8.2, the sequence of points on the SNS presenting the sequence of the Soft numbers C_k tends to the zero axis.
(2) $X_0 = 0$. In this case all points are on the 1 axis.
(3) $X_0 < 0$. The sequence of points presenting the Soft numbers C_k on the SNS does not exist.

It is to be noted that in both cases (1) and (2), the sequence of heights of the points on the SNS $A_k = X_k + Y_k = X_0 f_k'(Y_0) + f_k(Y_0)$ tends to infinity for any value of Y_0.

Conclusions The points in the dynamics of the Soft exponential function, defined on the SNS by the "convex" method, do exist if X_0 is a nonnegative real number. In this case, they always tend to infinity, either tending simultaneously to the zero axis ($X_0 > 0$) or being located on the 1 axis ($X_0 = 0$).

Example 8.8 Here we define a Soft function $f(C) = \cos(C)$, where $C = X\bar{0} \dotplus Y$ is any Soft number, as an extension of a real function $f(x) = \cos x$, by the following formula:

$$f(C) = \cos(C) = -X\sin(Y)\bar{0} \dotplus \cos(Y).$$

Given an initial Soft number C_0, this function recursively creates an infinite sequence of Soft numbers:

$$C_0 = X_0\bar{0} \dotplus Y_0,$$
$$C_1 = -X_0\sin(Y_0)\bar{0} \dotplus \cos(Y_0),$$
$$C_2 = X_0\sin(Y_0)\sin(\cos(Y_0))\bar{0} \dotplus \cos(\cos(Y_0)),$$

and so on. The continuation of the iterative process leads to the infinite sequence of Soft numbers $C_k = X_k\bar{0} \dotplus Y_k$, where $Y_k = \cos(\cos(\cos(\ldots(\cos(Y_0))))\ldots)$ with k names "cos" in the record, $X_k = (-1)^k X_0 \sin(Y_0)\sin(Y_1)\sin(Y_2)\ldots\sin(Y_{k-1})$, $k = 1, 2, 3, \ldots$.

It is proved in the real analysis that for any initial Y_0, the sequence Y_k tends to a constant irrational number $q = 0.739\ldots$. This fact can be verified numerically. If in a functional calculator any real number is chosen and then the key "cos" is pressed many times, the result is stabilized and shows the first digits of q.

If $X_0 = 0$ or $Y_0 = 0$, then $X_k = 0$ for all k. Otherwise, when k tends to infinity, X_k is a product containing an increasing number of factors with absolute values not exceeding $sin\ (0.8) < 0.8 < 1$, and therefore the values of X_k tend to zero.

In any case, the heights $A_k = X_k + Y_k$ of the points presenting the Soft numbers C_k on the SNS tend to $0 + q = q$, and therefore any "convex" point presentation of the sequence of Soft numbers created recursively by $f(\ C\ ) = \ cos\ (\ C\ )$ is bounded. If $X_0 = 0$ or $Y_0 = 0$, then such a sequence has a full point presentation on the SNS, resting on the 1 line. Otherwise the presentation is only partial, because the signs of X_k alternate and the condition $X_k Y_k > 0$ necessary for the "convex' point presentation of Soft numbers in the inner space of the SNS is satisfied only for a part of k (even or odd).

In any case, any "convex" point presentation on the SNS, full or partial, of any dynamic sequence of $cos(C\ )$ rests on the 1 line or tends to it, because the widths $B_k = \frac{Y_k}{X_k + Y_k}$ of the points are equal to 1 ($X_k = 0$) or tend to $\frac{q}{0+q} = 1$.

Resuming the above considerations, it is to be noted that any dynamic sequence created by $f(\ C\ ) = \ cos\ (\ C\ )$ on the SNS has a limit point $A = q$, $B = 1$.

8.3 Soft Dynamics Geometry

The Mandelbrot set (Mandelbrot 2004) is defined in the following way: For any complex number c, we define a sequence by a recursive instruction:

$$a_1(c) = c,$$
$$a_{n+1}(c) = (a_n(c))^2 + c,$$
$$(n = 1, 2, 3, \ldots).$$

The infinite sequence a_n(c) can be bounded or unbounded depending on the starting number c. The Mandelbrot set (Fig. 8.1) is defined as the set of all complex numbers c on the complex plane for which the sequence $a_n(c)$ is bounded.

We wished to investigate the dynamics of a similar function on the Soft number strip (SNS),$f(c) = c^2 + c$, where c is any Soft number. For this function, and with the help of a computer simulation, we obtained the following picture (Fig. 8.2), which is quite different from the picture for a complex c.

The notion of a "Soft dynamics set" of a given Soft function f, used in the description of Fig. 8.2 and of other figures further on in this section, refers to the set of Soft numbers c, classified according to the properties of the sequences created by the function f, by a recursion starting at c. Such sequences are called recursive or dynamic sequences of the function f, while all of them together constitute its dynamical system or dynamics.

The properties of dynamic sequences are connected to their point presentation on the SNS by the "convex method" and are marked by colors, as described in Table 8.1.

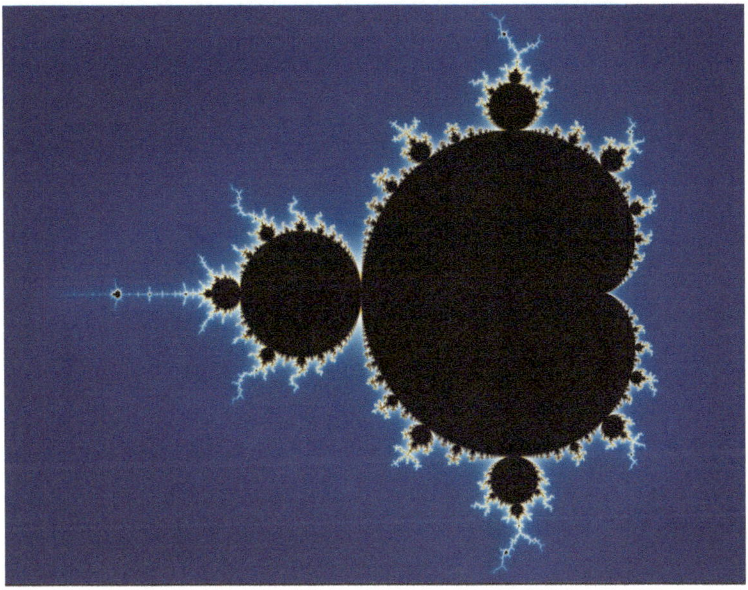

Fig. 8.1 The Mandelbrot set

Fig. 8.2 The Soft dynamics set of $f(c) = c^2 + c$

Table 8.1 The six domains of dynamics with Soft numbers

The sequence is …	Tending to the 0-line	Tending to the 1-line	Enclosed between the 0- and 1-lines
Bounded			
Not bounded			

Fig. 8.3 The Soft dynamics sets of $f(c) = c$

According to Table 8.1, any point in the red domain in Fig. 8.2 presents a Soft number that initiates a sequence of Soft numbers created recursively by the function $f(c) = c^2 + c$, for which a sequence of point images on the SNS found by the "convex" method is unbounded and tends to the zero line. A point in the green part of the picture has a similar meaning with one difference: the point sequence on the SNS is bounded, as is any point on the zero line. On the other hand, any point on the positive part of the right 1 line initiates an unbounded recursive sequence, which however does not tend to the zero line, remaining entirely on the 1 line, and therefore it must be yellow.

The following pictures show the Soft dynamics sets of some basic Soft functions, restricted to the positive right part of the SNS. A picture with a number of colors shows a number of sets corresponding to these colors. A one-color picture shows only one set.

The explanation of Fig. 8.3 is as follows: The Soft function $f(c) = c$ ("the identity function") has any Soft number as the start of a trajectory of a special, extreme kind: a fixed point. Thus, the dynamical system of this function coincides with the whole

Fig. 8.4 The Soft dynamics sets of $f(c) = c^2$

set of Soft numbers. The point convex presentations of some of them on the SNS fill the space between the zero line and the 1 line and these lines themselves. By the meaning of the colors stated in Table 8.1, the color of the points between these lines has to be white, their color on the zero line has to be green, and on the 1 line it is blue.

Clarifications for Fig. 8.4 may be found in Example 8.4.

The following pictures show the results of computer simulations and present the dynamics of several Soft functions defined as extensions of the corresponding real functions.

The special one-colored picture in Fig. 8.5 reflects the facts discovered in Example 8.8.

In Example 8.7, it is shown that all recursive sequences created by the Soft exponential function tend to infinity. In addition, they all tend to the zero axis, except those that started at a number on the 1 line and belong to it. All these facts are reflected in the picture shown in Fig. 8.6 The red color covers the whole area of the convex presentation on the SNS, except the 1 line, which has to be yellow.

The pictures of the Soft dynamic sets of some additional functions are shown in Figs. 8.7, 8.8 and 8.9 without an explanation of their forms.

(The zero line does not belong to the set, as the function is not defined on it.)

The Soft dynamics of the functions in Figs. 8.10 and 8.11 have not yet been studied theoretically. The pictures are a result of a computer experiment.

Fig. 8.5 The Soft dynamics set of $f(c) = \cos(c)$

Fig. 8.6 The Soft dynamics sets of $f(c) = e^c$

An experimental two-color picture. All sequences are bounded. The zero line is entirely green, and the 1 line is blue.

An experimental picture with a number of colors, including the colors of unbounded sequences, but not on the axes. The zero line is totally green, the 1 line is blue.

Fig. 8.7 The Soft dynamics sets of $f(c) = c^3$

Fig. 8.8 The Soft dynamics set of $f(c) = c^{0.5}$

8.4 Some Remarks Concerning the Results

The richness of appearance of the Soft dynamics sets opens one's mind toward a further investigation of the meaning of Soft dynamics. One of the topics in this direction is finding Soft dynamics sets for functions that cannot create an infinite recursive sequence starting at any Soft number in their domain. One such function is

Fig. 8.9 The Soft dynamics set of $f(c) = c + 1$

Fig. 8.10 The Soft dynamics sets of $f(c) = \sin(c)$

$\tan(c)$. This function, for example, is defined for $c = arctan\ (\pi/2)$, but already the second iteration of $\tan(c)$ does not exist for this value of c, and there is yet an infinite set of such numbers. Such a situation makes a principal distinction between Soft functions treated in this section and Soft tangents, so that investigating this function dynamics may be a real challenge.

In this chapter, we presented the richness of the geometric nature of the dynamics in a Soft number coordinate system. In the future, we aim to define the dimensions of Soft dynamics geometric shapes. In addition, we wish to extend the idea of a Soft number to a Soft operation. For example, the operation denoted @ could be both

Fig. 8.11 The Soft dynamics sets of $f(c) = c * \cos(c)$

addition or multiplication, so that $4 @ 6 = \{10, 24\}$, with a given probability that the operation will be addition or multiplication.

David Bohm believed in non-locality and searched for a mathematical theory to serve as the foundation for relativity theory and quantum theory. We wish to research the relevance of Soft logic for this purpose. Finally, we wish to research the technological aspect of Soft logic. Can we present the noise of certain electrical devices by using the multiplication of $\overline{0}$ with real numbers? Is the zero axis a mathematical presentation of nonlocal transmission? We hope to find answers, at least partial ones, to these questions.

References

Klein, M., Maimon, O.: The dynamics in the Soft numbers coordinate system. J. Adv. Math. **18**(1–17), 10.24297/jam.v18i.8531 (2020)

Mandelbrot, B.: Fractals and Chaos: The Mandelbrot Set and Beyond. Springer, New York (2004)

Chapter 9
Soft Analysis

This chapter discusses the development of Soft analysis. We define the distance between two Soft numbers and the absolute value of a Soft number. The innovation here lies in understanding that the distance between two points on the zero axis is a type of Soft zero. Using these properties, we define the basic neighborhood of Soft numbers and the unique typology of the set of Soft numbers and its mathematical properties.

9.1 Definition of the Distance in Any Set of Elements

The word "distance" is common in our daily lives. There is a specific distance between any two objects, whether on earth or in space, in our bodies or drawn on a piece of paper. All of these distances share the fact that they are measured using various units of measurement. Mathematics, with its ways of abstraction and generalization, offers a concept of measurement that does not rely on physical measurement and that can be applied even to the most abstract of objects, such as numbers. In the following two sections, we will discuss a mathematical approach to defining the distance between the elements of any set.

Definition 9.1 Let E be a set of elements w of any type. Let $d(w_1, w_2)$ be a function of every two elements in the set E with real values that fulfills the following conditions:

$$d(w_1, w_2) \geq 0,$$

$$d(w_1, w_2) = 0 \Leftrightarrow w_1 = w_2,$$

$$d(w_1, w_2) = d(w_2, w_1),$$

M. Klein, O. Maimon, *Foundations of Soft Logic*,
https://doi.org/10.1007/978-3-031-58233-2_9

$$d(w_1, w_2) \leq d(w_1, w_3) + d(w_3, w_2),$$

where w_1, w_2, w_3 are some elements in E.

The function $d(w_1, w_2)$ is called a distance function or *metric* of the set E, and the value d of every pair of elements (w_1, w_2) in E is called the distance between these two elements. We note immediately that according to this definition, if $d(w_1, w_2)$ is the metric of the set E, and $C > 0$ is some positive constant, then the function $C \cdot d(w_1, w_2)$ is also a metric in the set E. Thus, the same set E may have infinitely many metrics. Must they be proportional to each other? How can one define in practice the distance function in a given set E? The answers to these questions will be given further below.

Theorem 9.1 Let E be a set of any elements w that has an algebraic structure of an abelian group. In other words, in the set E, an addition operation between two elements is defined that follows the laws of associativity and commutativity and has a neutral element $w = 0$. Furthermore, every element w has an inverse element $-w$ such that $w + (-w) = 0$. In this set, a subtraction operation of elements can be defined according to the rule $w_1 - w_2 = w_1 + (-w_2)$.

Let $f(w)$ be a function defined for every element $w \in E$ with real values that fulfills the following conditions:

A. $f(w) \geq 0$ for every $w \epsilon E$.
B. $f(w) = 0 \Leftrightarrow w = 0$.
C. $f(-w) = f(w)$ for every $w \epsilon E$.
D. $f(w_1 + w_2) \leq f(w_1) + f(w_2)$,

where w_1, w_2 are some element in E.

Then the function of two elements $d(w_1, w_2) = f(w_1 - w_2)$ is a distance function in the set E and is called the distance function created by the function $f(w)$, and the function $f(w)$ is called the function that creates the distance function $d(w_1, w_2)$.

Proof 9.1 From the properties A–C of the function $f(w)$, it immediately follows that the function $d(w_1, w_2) = f(w_1 - w_2)$ fulfills the conditions A–C in the above definition. Let us check if condition D in the theorem leads to condition D in the distance function. Let w_1, w_2, w_3 be any elements in E. Then, according to the definition of the function $d(w_1, w_2) = f(w_1 - w_2)$ and the property D of the function f, we obtain the following:

$$\begin{aligned} d(w_1, w_2) &= f(w_1 - w_2) = f(w_1 - w_3 + w_3 - w_2) \\ &= f((w_1 - w_3) + (w_3 - w_2)) \leq f(w_1 - w_3) + f(w_3 - w_2) \\ &= d(w_1, w_3) + d(w_3, w_2) \end{aligned}$$

In other words, $d(w_1, w_2) \leq d(w_1, w_3) + d(w_3, w_2)$, where w_1, w_2, w_3 are any elements in E. Thus, the function $d(w_1, w_2) = f(w_1 - w_2)$ is a metric in E.

■

According to Theorem 9.1, it is possible to practically define a metric in the set E, if the set has an algebraic structure of an abelian group and we know the function $f(w)$ in the set E that fulfills the conditions A–D of this theorem.

As an example of such a description, we will examine the set $E = \mathbb{C} = \{z = x + y \cdot i | x, y \in \mathbb{R}, i^2 = -1\}$, which is the set of all complex numbers. With respect to the addition operation, this set has the algebraic structure of an abelian group. We will define within it the following function: $f(z) = f(x + y \cdot i) = C_1 \cdot |x| + C_2 \cdot |y| \quad (*)$, where C_1, C_2 are (some) positive constants, and for $r \in \mathbb{R}$ the absolute value of r is defined by the rule $|r| = \begin{cases} r, r \geq 0 \\ -r, r < 0 \end{cases}$.

It is easy to see that the function $f(z)$ defined in the formula (*) fulfills all the conditions A–D of the theorem, where $w = z$. In fact, the formula (*) describes infinitely many various functions $f(z)$ that create infinitely many metrics in the set $E = \mathbb{C}$:

$$d(z_1, z_2) = f(z_1 - z_2) = C_1 \cdot |x_1 - x_2| + C_2 \cdot |y_1 - y_2| \quad (**),$$

where $z_1 = x_1 + y_1 i$, $z_2 = x_2 + y_2 i$.

These metrics are not proportionate to each other. In order to select from them a specific metric, two additional independent conditions must be included, which will enable locating unequivocally the two constants C_1, C_2. For example, if we decide that $d(1, 0) = d(i, 0) = 1$, then we will obtain $C_1 = C_2 = 1$. Thus, the metric of the form (**), in which the number $z = 1$ and the number $z = i$ are at a distance of 1 from $z = 0$, is $d(z_1, z_2) = |x_1 - x_2| + |y_1 - y_2|$.

We note that the canonical metric in the set $\mathbb{C}$ that provides the same distance of 1 between $z = 1$, $z = i$, and $z = 0$, is $d(z_1, z_2) = \sqrt{(x_1 - x_2)^2 + (y_1 - y_2)^2}$

We now address the set $E = \mathbf{SN} = \{s = y\bar{0} \dotplus x | x \in \mathbb{R}, y \in \mathbb{R}, \bar{0} > 0, \bar{0}^2 = 0\}$. This set constitutes an abelian group with respect to addition operation of Soft elements. What are the metrics in the set **SN** and how many are there? We will answer these questions using a new theorem.

Theorem 9.2 Let E and F be any two sets, not necessarily groups, whose elements differ but are in a one-to-one correspondence: $\omega \in F \leftrightarrow w \in E$. Then, every metric $d(w_1, w_2)$ in the set E determines a metric $d(\omega_1, \omega_2)$ according to the following rule:

$$d(\omega_1, \omega_2) = d(w_1, w_2),$$

where

$$\omega_1 \in F \leftrightarrow w_1 \in E$$

$$\omega_2 \in F \leftrightarrow w_2 \in E$$

Proof 9.2 We check if the function of the two elements $d(\omega_1, \omega_2)$ defined in the set F fulfills all the conditions A–D for defining a metric. Let ω_1, ω_2, ω_3 be any elements in F, and let w_1, w_2, w_3 be their corresponding elements in E so that

$$w_1 \in E \leftrightarrow \omega_1 \in F$$

$$w_2 \in E \leftrightarrow \omega_2 \in F$$

$$w_3 \in E \leftrightarrow \omega_3 \in F$$

Consequently, based on the definition of $d(\omega_1, \omega_2)$ and on the fact that $d(w_1, w_2)$ is a metric in the set E, we obtain the following:

$$d(\omega_1, \omega_2) = (w_1, w_2) \geq 0,$$

$$d(\omega_1, \omega_2) = 0 \Leftrightarrow d(w_1, w_2) = 0 \Leftrightarrow w_1 = w_2 \Leftrightarrow \omega_1 = \omega_2,$$

$$d(\omega_1, \omega_2) = d(w_1, w_2) = d(w_2, w_1) = d(\omega_2, \omega_1),$$

$$\begin{aligned} d(\omega_1, \omega_2) &= d(w_1, w_2) \leq d(w_1, w_3) + d(w_3, w_2) \\ &= d(\omega_1, \omega_3) + d(\omega_3, \omega_2) \Rightarrow d(\omega_1, \omega_2) \leq d(\omega_1, \omega_3) + d(\omega_3, \omega_2) \end{aligned}$$

Thus, we have proved that the function of the two elements $d(\omega_1, \omega_2)$ in the set F fulfills all the conditions A–D for defining a metric in any given set. Therefore, $d(\omega_1, \omega_2)$, as defined in Theorem 9.2, constitutes a metric in the set F.

■

The elements of the set of complex numbers $\mathbb{C}$ and the elements of the set of Soft numbers **SN** have a one-to-one correspondence according to the rule $s = y\bar{0} \dotplus x \leftrightarrow z = x + yi$ $(x \in \mathbb{R}, y \in \mathbb{R})$.

Therefore, according to Theorem 9.2, every metric in set $\mathbb{C}$ that is expressed by real variables x, y becomes a metric in the set **SN**, where all these variables are viewed as properties of the Soft number $s = y\bar{0} \dotplus x$. Specifically, the infinitely many metrics (**) in the set $\mathbb{C}$ have various, parallel metrics in the set **SN**, which are defined according to the form: $d(s_1, s_2) = C_1|x_1 - x_2| + C_2|y_1 - y_2|$, where $s_1 = y_1\bar{0} \dotplus x_1, s_2 = y_2\bar{0} \dotplus x_2$, where $C_1 > 0$, $C_2 > 0$ are (any) two real, positive constants.

We emphasize that, as defined in the beginning of this section, every metric has real nonnegative values. Later, we will call the function $d(w_1, w_2)$—which is defined over the pairs of elements in the set E and whose values are real nonnegative—"a real metric in E." In a real metric, the distance between two elements in the set to which they belong is a real nonnegative number.

9.2 The Distance Between Two Soft Numbers

We note that the development of Sect. 9.1 regarding the determination of the distance between complex numbers and Soft numbers did not rely on a geometric presentation of these numbers. In this section, the theory will be developed further both geometrically and algebraically. For a geometric model of a set of Soft numbers, we define the Soft coordinate system. We will present this Soft coordinate system with several additional details related to the definition of a special metric in this space, called a Soft metric, in addition to the regular metric in this space, which is called a Euclidian metric.

In a Euclidian metric, the distance between two points M_1 and M_2 on the plane is defined as the length of the segment M_1M_2 that connects these two points, where the length is measured in relation to the same, sole, and single unit of length measurement selected in the Soft strip system when it was constructed.

The Euclidian metric in *the Soft coordinate system* has an algebraic expression $d(M_1, M_2) = \sqrt{(x_1 - x_2)^2 + (y_1 - y_2)^2}$, where x_1, y_1 are the pair of coordinates of the point M_1 in the *Soft system* and x_2, y_2 are the pair of coordinates of the point M_2 in the same system. Since Soft points and numbers have a one-to-one correspondence, then according to Theorem 9.2 (presented in Sect. 9.1), parallel to the Euclidian metric in the group of points in the Soft plane, there emerges a Euclidian metric in the set of Soft numbers **SN**:

$$d(s_1, s_2) = \sqrt{(x_1 - x_2)^2 + (y_1 - y_2)^2}, \text{where } s_1 = y_1\bar{0} \dotplus x_1, s_2 = y_2\bar{0} \dotplus x_2$$

For all intents and purposes, this metric is a metric in the set **SN**, a real metric as defined in the beginning of Sect. 9.1. However, this metric has a certain weakness as it cannot express the fundamental fact, which is related to Soft numbers, that the distance between the numbers $s_1 = \bar{0}$ and $s_2 = 0$ is smaller than the distance between any real number and zero. Following the definition of $\bar{0}$, we will denote the distance between $\bar{0}$ and absolute zero with the Greek letter τ (tau). Tau is a special number that has the following properties: $\tau > 0$, $\tau^2 = 0$.

Using τ, we will construct upon the Euclidian metric in the plane an additional metric that is not a real metric, which we will call a "Soft metric," denoted by two letters, *s.d.*, meaning "Soft distance." This metric is based on the assumption that $s.d.(\bar{0}, 0) = \tau$.

In the following we present, side by side, a definition of a Soft metric as a metric of a set of points on the Soft coordinate system and a definition as a metric in the set of Soft numbers represented by these points. We will also present the definition of the metric in a figure that contains three stages, alongside a definition of a metric in the Soft plane as a Soft distance between points.

Definition 9.2 A Soft metric in the set of Soft numbers SN. We define the following:

$$s.d.(\bar{0},0)=\tau,$$

(1) If $s_1=y_1\bar{0}\dotplus x_1, s_2=y_2\bar{0}\dotplus x_2$, and $y_1=y_2$, then s. d. $(s_1,s_1)=|x_1-x_2|$.
(2) If $s_1=y_1\bar{0}\dotplus x_1, s_2=y_2\bar{0}\dotplus x_2$, and $x_1=x_2$, then s. d. $(s_1,s_1)=|y_1-y_2|\cdot\tau$.
(3) If $s_1=y_1\bar{0}\dotplus x_1, s_2=y_2\bar{0}\dotplus x_2$, and $x_1\neq x_2$, $y_1\neq y_2$ then s. d. $(s_1,s_1)=|x_1-x_2|+|y_1-y_2|\cdot\tau$.

Comment 9.1 It is possible to merge the definition of a Soft metric into a single formula:

$$s.d.(s_1,s_2)=|x_1-x_2|+|y_1-y_2|\cdot\tau,$$

where $s_1=y_1\bar{0}\dotplus x_1, s_2=y_2\bar{0}\dotplus x_2$ (x_1, x_2, y_1, y_2 are any real numbers).

Comment 9.2 After defining the Soft metric, one naturally asks the question "What kind of numbers are the distances between Soft numbers, which are determined by this metric?" In response, we can say that these distances are dual nonnegative numbers or that they are "post-real" nonnegative numbers defined as real nonnegative numbers with an additional, special nonnegative number that is so small that its square is always zero.

9.3 The Neighborhood ε in the Set of Soft Numbers

In the previous section, we defined the distance between two Soft numbers $s_1=y_1\bar{0}\dotplus x_1, s_2=y_2\bar{0}\dotplus x_2$ in the following manner: $d(s_1,s_2)=|x_1-x_2|+|y_1-y_2|\tau$, where the *absolute value* of the real number is defined as usual, and the letter τ denotes the distance between $\bar{0}$ and 0: $=d(0,\bar{0})$.

As noted earlier, the difference between τ and $\bar{0}$ is that τ is an absolute positive constant that is smaller than any real positive number, such that $\tau^2=0$. In contrast to $\bar{0}$, it is not possible to present τ as a point on some axis; it can only be an object of thought, as it is entirely abstract.

The definition of a distance between Soft numbers goes hand in hand with the definition of the absolute value of a Soft number, so that $|s|=|y\bar{0}\dotplus x|=|x|+|y|\tau$, where $\tau=|\bar{0}|=|-\bar{0}|$. The relationship between the distance between Soft numbers and the absolute value in their region is $d(s_1,s_2)=|s_1-s_2|$.

We note that the classic definition of distance, or a metric, in a set of any elements assumes that the metric is a function of pairs of elements with real nonnegative values, which zeroes if and only if the elements in the pair are identical. The function does not depend on the order of the elements in the pair and fulfills the triangle inequality.

The metric defined above in the set of Soft numbers does not fulfill the first condition, which requires that its values be real numbers. The distance between Soft numbers is measured by a special two-part number, in which one of the parts is a real nonnegative number and the second part is a number that is not real. It can be characterized as a pre-real nonnegative number that is smaller than any real positive number. The structure of distances in the set of Soft numbers matches the structure of this set as a direct sum of two one-dimensional spaces: the space of real numbers, presented on the real axis, and the space of zero numbers, presented on the zero axis. Thus, from the point of view of linear algebra, the set of Soft numbers is a two-dimensional vector space. Defining a metric in this space as we did above turns this space into a special metric space that differs from regular metric spaces. We will now turn this space into a topological space in the pure classic sense.

In a set of some elements *M*, the topology of *T* is defined as a system of subsets of *M* that contains M and the empty set and that is union closed to any number of sets, whether finite or infinite, and intersection closed to a finite number of sets. For example, in a set of real numbers presented by a set of points on a real straight line, the topology T can be defined as a set of open intervals and all the sets obtained by a union, whether finite or infinite, of such intervals. This topology is Euclidian in the real straight line. It is possible to define a topology in the same set but in a different way, for example, the collection of all the rays (x, ∞), where x is any real number. It is easy to ascertain that the set of these subsets in *R* fulfills all the requirements of the topology *T*.

If the set *M* is a metric space, then we can use the metric to define the topology in *M*. This use is based on the concept of the neighborhood of each point in *M*, which is defined on the basis of the metric in *M*. We will follow this path in order to define the topology in the set of Soft numbers *SN*. We will use the special metric defined above in the set **SN**.

Definition 9.3 Let ε be a real positive number, as small as we wish. Let $s_0 = y_0\bar{0} \dotplus x_0$ be any constant Soft number. The subset of Soft numbers $s = y\bar{0} + x$ whose distance from s_0 is $d(s, s_0) = |x - x_0| + |y - y_0|\tau$ and that fulfills the conditions $|x - x_0| < \varepsilon$ as well as $|y - y_0| < \varepsilon$ is called "the neighborhood ε" of s_0. $Re(s)$ denotes the real part of the Soft number s and that $Zec(s)$ is the real coefficient of $\bar{0}$ in the structure of s. We then use $\omega(s, \varepsilon)$ to denote the neighborhood ε of the Soft number s.

According to Definition 9.1, there is

$$\omega(s_0, \varepsilon) = \{s \in \mathbf{SN} \mid |Re(s) - Re(s_0)| < \varepsilon, |Zec(s) - Zec(s_0)| < \varepsilon \} \text{ for } \varepsilon > 0.$$

From this definition, it follows that if the formula $0 < \varepsilon_1 < \varepsilon_2$, then $\omega(s, \varepsilon_1) \subseteq \omega(s, \varepsilon_2)$ for every $s \in \mathbf{SN}$.

Definition 9.4 Let Ω be any subset in the set of Soft numbers **SN**. We will call the subset Ω an open set in the following two cases:

(a) Ω is an empty set.
(b) Ω is not an empty set, and for every Soft number s that belongs to Ω, there is a neighborhood that belongs entirely to Ω. In other words, $s \in \Omega \Rightarrow \omega(s, \varepsilon) \subseteq \Omega$ for any number $0 < \varepsilon$.

For example, the subset of Soft numbers s for which $Re\,(s) + Zec(s) < 0$ is an open set, and the set of all the Soft numbers s for which $Re(s) + Zec(s) = 0$ is a non-open set.

Here we note only that according to Definition 9.4, the set of all the Soft numbers SN is an open set. Within the overall set SN, there are infinitely many open sets of Soft numbers. It is possible to prove that the collection of all the open sets of Soft numbers is closed under union of any number of such sets, finite or infinite, and closed under the intersection of a finite number of open sets. Thus, the collection of all the open sets of Soft numbers constitutes a topology in the space SN of Soft numbers. Further along, we will see that this topology is presented on the Soft coordinate system by the Euclidian topology. We will start by presenting the neighborhood $\omega(s, \varepsilon)$ in the strip of Soft numbers.

In a Soft coordinate system with two axes $x = Re\,(s)$ and $y = Zec(s)$, every point (x, y) presents the Soft number $s = y\bar{0} \dotplus x$. Therefore, every set of Soft numbers is presented by a set of points in the Soft strip. In particular, the neighborhood ε of the Soft number $s = y\bar{0} \dotplus x$ is presented by the set of points within the neighborhood that has as its center the point presenting the number s. Every open set of Soft numbers is presented in the Soft plane by the set of points that is called "the open set in the plane."

The topology defined in the set of Soft numbers is presented in the Soft plane by a collection of open sets in the plane that constitute the Euclidian topology in the plane. This topology goes hand in hand with the Euclidian topology in the plane. On the other hand, the topology defined above in the set of Soft numbers follows the definition of the special metric in this set, but it is not actually based on this metric. Herein lies the uniqueness of Soft analysis compared to the real analysis of functions with two variables and the complex analysis of complex functions. These two regions use the Euclidian metric and the Euclidian topology in the plane, while Soft analysis uses both the Euclidian topology in the plane and the special metric, which is not a Euclidian metric in the plane. This duality means that in some cases, Soft analysis and traditional analyses share a commonality, and in others there is a fundamental difference. This situation will be manifested in the next chapter, which is dedicated to the concept of limits.

We conclude this chapter with Fig. 9.1, which presents the neighborhood $\varepsilon = 0.5$ of the number $s = 3\bar{0} \dotplus 2$ in the Soft plane and the open set of Soft numbers s with $Re(s) + Zec(s) < 0$, where several points from this group are marked with the square neighborhoods belonging to the set.

For the presentation of the neighborhood of s in the Soft coordinate system, we will use the formula from Chap. 5.

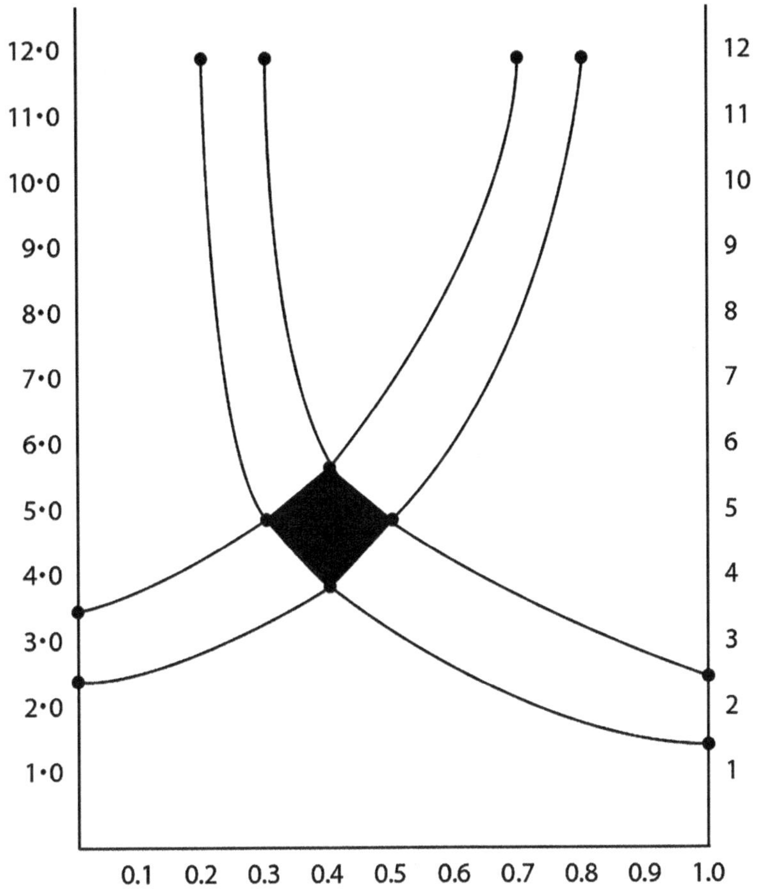

Fig. 9.1 An open set of Soft numbers

$$s = x\bar{0} \dotplus y$$

$$A = (|x| + |y|)\text{sign}(x)$$

$$B = \frac{y \cdot \text{sign}(x)}{|x| + |y|}$$

$$s = 3\bar{0} \dotplus 2$$

$$x = 3, \quad y = 2$$

$$A = 5$$

$$B = \frac{2}{5} = 0.4$$

Table 9.1 Calculation of the neighborhood

B	1.5/*B*	2.5/*B*	2.5/(1 – *B*)	3.5/(1 – *B*)
0			2.5	3.5
0.1	15	25	2.777778	3.888889
0.2	7.5	12.5	3.125	4.375
0.3	5	8.333333	3.571429	5
0.4	3.75	6.25	4.166667	5.833333
0.5	3	5	5	7
0.6	2.5	4.166667	6.25	8.75
0.7	2.142857	3.571429	8.333333	11.66667
0.8	1.875	3.125	12.5	17.5
0.9	1.666667	2.777778	25	35
1	1.5	2.5		

For the simplicity of the calculation, we can assume that in the neighborhood $x, y > 0$,

$$A = x + y$$

$$B = \frac{y}{x + y}$$

$$y = AB$$

$$x = A - y = A - AB = A(1 - B)$$

$$\mathrm{U} = \left\{x\bar{0} \dotplus y \,\middle|\, |x - 3| < 0.5 \text{ and } |y - 2| < 0.5\right\}$$

$$|A(1 - B) - 3| < 0.5$$

$$2.5 < A(1 - B) < 3.5$$

$$|AB - 2| < 0.5$$

$$1.5 < AB < 2.5$$

Results of calculations appear in Table 9.1 and are present in Fig. 9.1

Chapter 10
Limits of Soft Numbers

This chapter deals with the limits of Soft numbers. In Soft analysis, as in real and complex analysis, limits are the foundation and primary tool for the investigation and description of the advanced properties of functions. In the first section of this chapter, we define the concepts of finite and infinite limits of an infinite sequence of Soft numbers, as well as the related concepts of convergent and divergent sequences and bounded and unbounded sequences. We investigate the relationship between convergent and bounded sequences of Soft numbers, and then, using the limits of sequences, we define the concept of the limit of a Soft function at a point. We also investigate and present the algebraic properties of limits in the Soft numbers region.

10.1 The Limit of an Infinite Sequence of Soft Numbers

Let us denote by $\{s_n\}$ an infinite set of Soft numbers $s_n = y_n\bar{0}\dot{+}x_n$, where x_n, y_n are real numbers and n is a natural number $\boldsymbol{n} = \mathbf{1,2,3,} \ldots$

Definition 10.1 A finite limit of an infinite sequence

Let $\{s_n\}$ be any infinite sequence of Soft numbers. The constant Soft number s_0 is called the limit of the sequence where n tends to infinity, if and only if for every number $\epsilon > 0$ there exists a natural number N that depends on ϵ ($\mathrm{N} = N_\varepsilon$) such that all the elements of the sequence $\{s_n\}$ with the numbers $n > N_\varepsilon$ are in the neighborhood ϵ of the same constant number s_0. Such a case we denote by $s_0 = \lim\limits_{n\to\infty} s_n$.

Theorem 10.1 The Soft number s_0 is the finite limit of the infinite sequence of Soft numbers s_n if and only if there exists $y_0 = \lim\limits_{n\to\infty} y_n$,$x_0 = \lim\limits_{n\to\infty} x_n$, where $x_0 = \mathrm{Re}(s_0)$, $y_0 = \mathrm{Zec}(s_0)$, $x_n = \mathrm{Re}(s_n)$, $y_n = \mathrm{Zec}(s_n)$ $n = 1, 2, 3, \ldots.$

M. Klein, O. Maimon, *Foundations of Soft Logic*,
https://doi.org/10.1007/978-3-031-58233-2_10

Proof 10.1 From Definitions 9.1 and 9.3 of the neighborhood ϵ in Sect. 9.3, it follows that $\lim_{n\to\infty} s_n = s_0$ if and only if for every ϵ > 0 there exists $N = N(\epsilon)$ such that $|x_n - x_0| < \varepsilon, \quad |y_n - y_o| < \varepsilon, n > N(\varepsilon)$ (**), where $x_n =$ Re (s_n), $y_n = \text{Zec}(s_n)$, $n = 1$, 2, 3.... By the definition of the limit of a real sequence, the conditions (**) are equivalent to the conditions $\lim_{n\to\infty} x_n = x_0$, $\lim_{n\to\infty} y_n = y_{0.}$ From this follows the theorem.

■

Example 10.1 Given a sequence of Soft numbers $s_n = \frac{1}{1+2^{-n}}\bar{0} \dot{+} \left(-\frac{n+1}{n}\right)$, $n = 1, 2, 3, \ldots$. In this case, $x_n = \text{Re}(s_n) = \frac{n}{n+1}$, $y_n = \text{Zec}(s_n) = \frac{1}{1+2^{-n}}$, and there exists $\lim_{n\to\infty} x_n = \lim_{n\to\infty} y_n = 1$. It follows that according to Theorem 10.1, we conclude that $\lim_{n\to\infty} s_n = 1 \dot{+} \bar{0}$.

Example 10.2 Let $s_n = (-1)^n \bar{0} \dot{+} e^{-n}$; $n = 1, 2, 3, \ldots$. In this case, $x_n =$ Re (s_n) $= e^{-n}$, $\text{Zec}(s_n) = (-1)^n$, where $n = 1, 2, 3, \ldots$. The limit $\lim_{n\to\infty} e^{-n}$ exists and equals zero, and the limit $\lim_{n\to\infty} y_n = \lim_{n\to\infty} (-1)^n$ does not exist. According to Theorem 10.1, the sequence of Soft numbers given in Example 10.2 has no limit in the region of Soft numbers.

So far, we have defined and demonstrated the limit of a sequence of Soft numbers that is itself a Soft number. Such a limit is called a finite limit, and it is presented by a regular point in the Soft strip to which the points that present the elements of the sequence tend. Now we will move on to the definition of an infinite limit of a sequence of Soft numbers. We recall that in the region of real numbers, there are two infinite limits, $-\infty$ and $(+\infty)$. In the region of complex numbers, there is only one infinite limit, infinity, which is defined by a positive infinite limit of a sequence of positive real numbers as follows: $\lim_{n\to\infty} z_n = \infty \leftrightarrow \lim_{n\to\infty} |z_n| = \infty$, where the absolute value of the complex number z is defined by the formula $|z| = \sqrt{|\text{Re}(z)|^2 + |Im(z)|^2}$.

In defining infinite limits in the region of Soft numbers, we will adopt the existing approach in the region of complex numbers. In other words, we define $\lim_{n\to\infty} s_n = \infty \leftrightarrow \lim_{n\to\infty} |s_n| = \infty$, where $\{\boldsymbol{s_n}\}$ is a sequence of some Soft numbers and according to the definition in Chap. 9, $|s| = |\text{Re}(s)| + |\text{Zec}(s)|\tau$.

In this formula, s is some Soft number, and τ is a pre-real positive number smaller than any real positive number; in other words, $0 < \tau < \varepsilon \quad \forall \varepsilon > 0 \in R$. From this definition of the absolute value of a Soft number, it follows that $|y|\tau < \varepsilon$ for every real number y and for every real number $\varepsilon > 0$. From this fact follows the equivalence

$$\lim_{n\to\infty} |s_n| = \lim_{n\to\infty} (|\text{Re}(s_n)| + |\text{Zec}(s_n)|\tau) = \infty \leftrightarrow \lim_{n\to\infty} |\text{Re}(s_n)| = \infty$$

All these considerations lead to the following definition.

Definition 10.2 Let $\{s_n\}$ be an infinite sequence of Soft numbers. We say that this sequence has an infinite limit denoted by $\underset{n\to\infty}{\text{Lim}}\, s_n = \infty$, if and only if there exists $\lim_{n\to\infty} |\text{Re}(s_n)| = \infty$. In other words, for every real number M, $M > 0$, there exists a natural number N that depends on M such that there exists $|\text{Re}(s_n)| > M$ for every $n > N$.

Remark 10.1 According to Definition 10.2, the existence of an infinite limit of a sequence of Soft numbers depends on the behavior of the sequence of the real parts of its elements; it is not affected by the behavior of its null parts. For example, for the sequence of Soft numbers $s_n = y_n\bar{0} \dot{+} (-1)^n\sqrt{n}, y_n \in R, n = 1, 2, 3, \ldots,$ there exists an infinite limit as long as the real numbers are not $\boldsymbol{y_n}$, since $\lim_{n\to\infty} |\text{Re}(s_n)| = \lim_{n\to\infty} \sqrt{n} = \infty$.

Comment 10.1 If all the null parts of the elements of the sequence $\{s_n\}$ equal zero, this sequence becomes a sequence of real numbers. Notably and interestingly, in this case, the existence of an infinite limit for this sequence depends on the set of overall numbers it is viewed in. For example, let us examine the sequence of numbers $(-1)^n 2^n$, where $n = 1, 2, 3, \ldots$. If we view this sequence as a sequence of Soft numbers, in other words we locate it within the set **SN**, then it has an infinite limit because

$$\lim_{n\to\infty} |(-1)^n 2^n| = \lim_{n\to\infty} 2^n = \infty$$

If we locate this sequence within the overall set of real numbers R, then it has no limit at all, and especially no infinite limit, because the elements of the sequence tend to $+\infty$ when n is even and to $-\infty$ when n is odd.

Comment 10.2 If in the sequence (s_n), the real parts of all the elements of the sequence equal zero, then the result is a sequence of null numbers $\{y_n\bar{0}\}$,$y_n \in R$, $n = 1, 2, 3, \ldots$. Such a sequence cannot have an infinite limit because $\lim_{n\to\infty} |\text{Re}(s_n)| = 0$. The finite limit of the infinite sequence $\{y_n\bar{0}\}$ exists if and only if the sequence of real numbers $\{y_n\}$ has a finite limit. For example, the sequence of null numbers $\frac{1}{n}\bar{0}$,$n = 1, 2, 3, \ldots$ has a finite limit, which is absolute zero:

$$\lim_{n\to\infty} \frac{1}{n}\bar{0} = \left(\lim_{n\to\infty} \frac{1}{n}\right)\bar{0} = 0\bar{0} = 0$$

To conclude this section, we will refer to the concept of a bounded sequence. The infinite sequence of Soft numbers $\{s_n\}$ is called a *bounded sequence* if and only if there is a real positive number C such that $|s_n| < C$, $n = 1, 2, 3, \ldots$.

This definition corresponds to the definition of a bounded sequence in real analysis and complex analysis. In both of these traditional methods of analysis, there is a theorem according to which in every bounded sequence there exists a subsequence with a finite limit. This argument is not valid in Soft analysis. For example, the infinite sequence of Soft numbers $\{n\bar{0}\}$ is bounded by every natural n, and every real number $\varepsilon > 0$ fulfills $|n\bar{0}| = n\tau < \varepsilon$. Nonetheless, within the infinite sequence $\{n\bar{0}\}$, there is no infinite subsequence with a finite limit because the entire sequence of natural numbers and every subsequence within it tend to $+\infty$. Notably, every situation in which there is a difference between Soft analysis and the traditional analyses is a result of the unique properties of the number $\bar{\mathbf{0}}$, which has a major and crucial role in the creation of the set of Soft numbers that is the foundation of Soft analysis.

10.2 The Limit of a Soft Function at a Point

The concept of a Soft function was defined and examined from an algebraic point of view in Chap. 6. The shift to a topological view of this subject requires, as a first step, defining the concept of the limit of a Soft function at a point.

Definition 10.3 Let $f(s)$ be a Soft function of a Soft variable s with a domain $D(f) \subseteq SN$. Let s_0 be a constant Soft number in the domain of $f(s)$ or at the limit of this domain. We denote an infinite sequence of Soft numbers by $\{s_n\}$. The sequence accumulates to the number s_0 if and only if all the various elements of the sequence differ from s_0 and tend to s_0; in other words, $\lim_{n\to\infty} s_n = s_0$. We say that the function $f(s)$ has a finite/infinite limit denoted by $\lim_{s\to s_0} f(s)$ at the point s_0, if and only if for every infinite sequence of Soft numbers $\{s_n\}$ that belongs in its entirety to $D(f)$ and accumulates to s_0, the infinite sequence of matching values of the function $f(s)$ tends to the *same finite/infinite limit*. This limit is accepted as the limit of the function $f(s)$ at the point s_0.

In other words, $\lim_{s\to s_0} f(s) = \lim_{n\to\infty} f(s_n)$ $(*)$, where $\{s_n\}$ is some infinite sequence in the domain of $f(s)$ that accumulates to s_0 on the condition that there exists a finite/infinite limit on the right side of the equality (*) and it is not dependent on the selection of the sequence $\{s_n\}$.

This definition has an obviously theoretical character. The tools to make a practical calculation of the limits of Soft functions will be developed in the future. In anticipation of this development and according to the definition given above, we note that the concept of "the limit of the Soft function $f(s)$ at the point s_0" is interpreted as follows: "the limit of the Soft function $f(s)$, where the Soft variable s tends to the constant Soft number s_0 while remaining different from it." This interpretation is reflected in the notation $s \to s_0$, which shows that the variable s tends to s_0, while it is at the other end of the process, meaning $s \neq s_0$.

The following theorem presents the limits of two Soft functions—the constant function and the identity function.

Theorem 10.2

(A) Let $f(s)$ be a constant Soft function, meaning $f(s) = c, \quad s \in SN$, where c is any given Soft constant number. Let s_0 be some constant Soft number, then there exists a finite limit of $f(s)$ at the point s_0 that equals c: $\lim_{s \to s_0} f(s) = c$.

(B) Let $f(s)$ be a Soft identity function, in other words, $f(s) = s, \ s \in SN$. Let s_0 be some constant Soft number, then there exists a finite limit of $f(s)$ at the point s_0 that equals s_0; in other words, $\lim_{s \to s_0} f(s) = s_0$.

The next theorem refers to the limits of any Soft function and is based on presenting them as a combination of real functions of two variables, where the vector of the coefficients of the combination is $(1, \bar{0})$.

Theorem 10.3 Let $f(s)$ be a Soft function of the Soft variable s, if it is in the domain $D(f) \subseteq SN$ given by the form $f(s) = v(x,y)\bar{0} \dot{+} u(x,y)$ (**), where $x =$ Re (s), $y =$ Zec(s) $v(x,y)$, $u(x,y)$ are two real functions defined in the same domain that presents the domain $D(f)$ in the Soft strip. Let s_0 be a constant Soft number in the domain of $f(s)$ or at its limits, and let (y_0, x_0) be the point that presents s_0 on the Soft plane. With these notations, the following arguments are valid:

(A) The finite limit of the function $f(s)$ at the point s_0 exists if and only if the functions $u(x,y)$ and $v(x,y)$ have finite limits at the point (y_0, x_0), which are defined by their definition in the real analysis of the functions of two variables. In such a case there exists

$$\lim_{s \to s_0} f(s) = \left(\lim_{\substack{x \to x_0 \\ y \to y_0}} v(x,y) \right) \bar{0} \dot{+} \lim_{\substack{x \to x_0 \\ y \to y_0}} u(x,y)$$

(B) The infinite limit of $f(s)$ at the point s_0 exists if and only if the real function $|u(x,y)|$ has an infinite limit at the point (y_0, x_0). In other words, $\lim_{s \to s_0} f(s) = \infty \Leftrightarrow \lim_{\substack{x \to x_0 \\ y \to y_0}} |u(x,y)| = \infty$.

If neither of the cases A nor B exists, then the function $f(s)$ does not have a limit, finite or infinite, at the point s_0. The arguments of the theorem follow from the definitions in this section and the previous one and from the definitions and properties of the limits of real functions.

■

Theorem 10.3 enables implementing the calculation of the limits of Soft functions to the calculation of real functions, thus using the techniques that were developed for this purpose in Soft analysis.

Comment 10.3 In special cases, it is possible that in the form (**), the function $u(x, y)$, the function $v(x, y)$, or even both are in fact dependent on a single variable, either y or x. In such cases, the limit of the function of two variables at the point (x_0, y_0) becomes the limit of the function of one variable, either y or x, at the corresponding point x_0 or y_0.

Example 10.3 Let $f(s)$ be a Soft function given by the formula: $f(s) = \frac{\sin(y-1)}{y-1}\bar{0}\dot{+}e^{\frac{x}{(y-1)^2}}$, where $x =$ Re (s); $y = \mathrm{Zec}(s)$.

This formula defines a function of the Soft variable s, although we cannot write a closed algebraic expression with relation to s for this function. The domain of this function is the set of all Soft numbers s with Zec$(s) \neq 1$. If x_0 is some real constant number, then the number $s_0 = \bar{0}\dot{+}x_0$ is in the domain of the given function $f(s)$, and we may question the existence and limit of $f(s)$ at the point s_0. To answer these questions, we must find the limits of the following real functions:

$$\lim_{\substack{x\to x_0\\ y\to 1}} \frac{x}{(y-1)^2} = \begin{cases} -\infty\ , & x_0 < 0 \\ \infty\ , & x_0 > 0 \\ \text{does not exist} & x_0 = 0 \end{cases}$$

$$\Rightarrow \lim_{\substack{x\to x_0\\ y\to 1}} e^{\frac{x}{(y-1)^2}} = \begin{cases} 0, & x_0 < 0 \\ \infty & , x_0 > 0 \\ \text{does not exist} & x_0 = 0 \end{cases}$$

and also

$$\lim_{y\to 1} \frac{sin(y-1)}{y-1} = \lim_{t\to 0} \frac{\sin(t)}{t} = 1.$$

Based on these results and according to Theorem 10.3, we conclude that

$$\lim_{s\to s_0 \ \ s_0 = \bar{0}\dot{+}x_0} f(s) = \begin{cases} \bar{0}, & x_0 < 0 \\ \infty, & x_0 > 0 \\ \text{not exist}, & x_0 = 0 \end{cases}$$

10.3 The Properties of the Limits of Soft Functions

If the intersection of the domains of two Soft functions is non-empty, then four operations can be applied to them: addition, subtraction, multiplication, and division. The following theorem argues that under certain conditions, algebraic operations with Soft functions are accompanied by similar operations at their limits.

Theorem 10.4 Let $f(s)$, $g(s)$ be Soft functions with the domains $D(f)$, $D(g)$, respectively, that have a non-empty intersection $D = D(f) \cap D(g) \neq \emptyset$. Let s_0 be some constant Soft number in the domain D or at its limit. If each of the functions $f(s)$, $g(s)$ has a finite limit at the point s_0, then each of the sum, difference, and product functions has a finite limit at the point s_0, and there exists

$$\lim_{s \to s_0} (f(s) + g(s)) = \lim_{s \to s_0} f(s) + \lim_{s \to s_0} g(s),$$

$$\lim_{s \to s_0} (f(s) - g(s)) = \lim_{s \to s_0} f(s) - \lim_{s \to s_0} g(s),$$

$$\lim_{s \to s_0} (f(s) \cdot g(s)) = \left(\lim_{s \to s_0} f(s) \right) \cdot \left(\lim_{s \to s_0} g(s) \right),$$

In addition, if the limit of $g(s)$ at the point s_0 is not a null number, then there exists a finite limit of the quotient function $\frac{f(s)}{g(s)}$ at the point s_0, and there exists $\lim_{s \to s_0} \frac{f(s)}{g(s)} = \frac{\lim_{s \to 0} f(s)}{\lim_{s \to s_0} g(s)}$.

This theorem follows from the operation rules of Soft functions and from the rules of the limits of real functions.

■

Theorem 10.5 Let $f(s)$ be a Soft rational function, meaning $f(s) = \frac{P(s)}{Q(s)}$, where $P(s)$ and $Q(s)$ are Soft polynomials. Let s_0 be a constant Soft number such that the value of the polynomial $Q(s)$ for $s = s_0$ is not a null number, meaning that $Q(s_0) \neq r\bar{0}$ for every $r \in R$. Then the function $f(s)$ has a finite limit at the point s_0, and there exists: $\lim_{s \to s_0} f(s) = \frac{P(s_0)}{Q(s_0)}$.

Proof 10.5 From Theorem 10.3 in the previous section and Theorem 10.4 in this section, it follows that for every Soft number s_0, every natural exponent n, and every constant Soft coefficient c, there exists $\lim_{s \to s_0} (cs^n) = cs_0^n$. From this equality, from the definition of Soft polynomials, and from Theorem 10.4, it follows that the following finite limits exist: $\lim_{s \to s_0} P(s) = P(s_0)$, $\lim_{s \to s_0} Q(s) = Q(s_0)$.

From these equalities and from the arguments of Theorem 10.4 that refer to division operations, it follows that if $Q(s_0)$ is not a null number $r\bar{0}$, $r \in R$, then there exists a finite limit of $f(s)$ at the point s_0, and there exists $\lim_{s \to s_0} f(s) = \frac{P(s_0)}{Q(s_0)}$.

■

Example 10.4 Let us observe the rational function $f(s) = \frac{1 \cdot \bar{0} + s}{s^2 - 1}$, which is the quotient of two Soft polynomials: the first-degree polynomial $P(s) = 1\bar{0} + s$ and the second-degree polynomial $Q(s) = s^2 - 1$. The polynomial $Q(s)$ has null values if and only if $Re(s) = \pm 1$. For every constant Soft number s_0 with $Re(s) \neq \pm 1$, the function $f(s)$ has a finite limit at the point s_0, and there exists $\lim_{s \to s_0} f(s) = \frac{P(s_0)}{Q(s_0)} = \frac{1 \cdot \bar{0} + s_0}{s_0^2 - 1}$. In particular, if s_0 is a Soft null number, meaning that $s_0 = r\bar{0}, r \in R$, then there exists $\lim_{s \to s_0 = r\bar{0}} f(s) = \frac{r\bar{0} + 1\bar{0}}{(r\bar{0})^2 - 1} = \frac{(r+1)\bar{0}}{0 - 1} = -(r + 1)\bar{0}$.

Chapter 11
The Soft Möbius Function and the Riemann Hypothesis

The Riemann hypothesis is the conjecture that the Riemann zeta function has its zeros only at the negative even integers and complex numbers with a real part $\frac{1}{2}$. It is considered the most important unsolved problem in pure mathematics, and it is known that there is a deep connection between the Riemann hypothesis and the Möbius function of number theory (Edwards 2001).

This chapter extends the Möbius function to the Soft Möbius function. The regular Möbius function, which is related to prime factorization, accepts three values: 1, 0, and -1. Using Soft logic, we extend the function to the following five values: $-1, -0, 0, +0,$ and 1. We prove the basic properties of the Soft Möbius function and discuss the properties of the Soft Mertens function, which is the aggregate sum of the values of the Soft Möbius function. We show that the relationship between the properties of the Mertens function and the Riemann hypothesis is correct also when applied to the relationship between the Soft Mertens function and the Riemann hypothesis.

11.1 The Definition of the Soft Möbius Function

The fundamental theorem of arithmetic states that every natural number n can be presented in a unique way as a product of powers of prime numbers by the equation $n = {p_1}^{\alpha_1} \cdot {p_2}^{\alpha_2} \cdot \ldots {p_r}^{\alpha_r}$. The index of factorization $k(n)$ is defined by $k(n) = \sum_{i=1}^{r} \alpha_i$.

Let us define the following set:

$$T_0 = \{1\}$$

and

M. Klein, O. Maimon, *Foundations of Soft Logic*,
https://doi.org/10.1007/978-3-031-58233-2_11

$$T_1 = \{n \ : n > 1, \ \ \alpha_1 = \alpha_2 = .. = \alpha_r = 1\}$$

The regular Möbius function in number theory is defined by

$$\mu(n) = \begin{cases} 1 & n \in T_0 \\ (-1)^{k(n)} & n \in T_1 \\ 0 & else \end{cases}$$

and the Mertens function is defined by

$$\mathrm{M}(x) = \sum_{n \leq x} \mu(n)$$

The Riemann hypothesis is equivalent to that for every $\varepsilon > 0$

$$\lim_{x \to \infty} \frac{M(x)}{x^{\frac{1}{2}+\varepsilon}} = 0$$

The Spencer-Brown Formula for $\mu(n)$ is

$$\mu(n) = 1 - n - \sum_{d=2}^{n-1} \mu(d) \left[\frac{n}{d}\right]$$

The Legendre/Spencer Brown formula is

$$\pi(n) = \pi\left(\sqrt{n}\right) + \sum_{d=1}^{\sqrt{n}} \mu(d) \left[\frac{n}{d}\right]$$

For the definition of the Soft Möbius function, we split the set of natural numbers N into four sets:

$$T_0 = \{1\}$$

$$T_1 = \{n \ : n > 1, \ \ \alpha_1 = \alpha_2 = \ldots = \alpha_r = 1\}$$

$$T_2 = \{n \ : \ \ \alpha_1 = \alpha_2 = \ldots = \alpha_{q-1} = \alpha_{q+1} = \ldots = \alpha_r = 1 \text{ and } \alpha_q = 2 \text{ or } \alpha_q = 3\}$$

$$T_3 = N \backslash (T_0 \cup T_1 \cup T_2)$$

We define the Soft Möbius function $\tilde{\mu}(1) = 1$ for $n > 1$:

$$\tilde{\mu}(n) = \begin{cases} (-1)^{k(n)} & n \in T_1 \\ (-1)^{k(n)}\overline{0} & n \in T_2 \\ 0 & n \in T_3 \end{cases}$$

Theorem 11.1 Let π be the set of prime numbers. Then $T_2 = \bigcup_{p\in\pi} p^2 \cdot \{1, T_1\}$.

Proof 11.1 Let us define T so that

$$T = \bigcup_{p\in\pi} p^2 \cdot \{1, T_1\}$$

$$m \in T \rightarrow m = p^2 \quad \text{or} \quad m = p^2 l, \quad l \in T_1$$

and therefore $m \in T_2$.

Now, if $m \in T_2$, then $m = p^2$ *or* $m = p^2 l \quad l \in T_1$, then $m \in T$. This proves that

$$T_2 = T = \bigcup_{p\in\pi} p^2 \cdot \{1, T_1\}$$

■

11.2 The Sum of the Soft Möbius Function

$$\tilde{M}(x) = \sum_{n\leq x} \tilde{\mu}(n) = S(x)\bar{0}\dot{+}M(x)$$

We present some calculations in Table 11.1

Let us denote the probabilities of -1, 1, 0, $-$ 0, + 0 by $p_{-1}, p_1, p_0, p_{-0}, p_{+0}$.

Remark 11.1 It is known from number theory that $p_{-1} + p_1 = \frac{6}{\pi^2}$.

Theorem 11.2

$$S(x) = \sum_{p\leq\sqrt{x}} M\left(\left[\frac{x}{p^2}\right]\right)$$

Proof 11.2

$$M(x) = 1 + \sum_{j\in T_1, j\leq x} (-1)^{k(j)}$$

$$S(x) = \sum_{j\in T_2, j\leq x} (-1)^{k(j)}$$

According to Theorem 11.1,

Table 11.1 The sum of the Soft Möbius function

n	$\mu(n)$	$\tilde{\mu}(n)$	M(n)	$S(n)$	$\tilde{M}(n)$
1	1	1	1	0	$0\bar{0}\dot{+}1$
2	−1	−1	0	0	$0\bar{0}\dot{+}0$
3	−1	−1	−1	0	$0\bar{0}\dot{+}(-1)$
4	0	0+	−1	1	$1\bar{0}\dot{+}(-1)$
5	−1	−1	−2	1	$1\bar{0}\dot{+}(-2)$
6	1	1	−1	1	$1\bar{0}\dot{+}(-1)$
7	−1	−1	−2	1	$1\bar{0}\dot{+}(-2)$
8	0	−0	−2	0	$0\bar{0}\dot{+}(-2)$
9	0	+0	−2	1	$1\bar{0}\dot{+}(-2)$
10	1	1	−1	1	$1\bar{0}\dot{+}(-1)$
11	−1	−1	−2	1	$1\bar{0}\dot{+}(-2)$
12	0	−0	−2	0	$0\bar{0}\dot{+}(-2)$
13	−1	−1	−3	0	$0\bar{0}\dot{+}(-3)$
14	1	1	−2	0	$0\bar{0}\dot{+}(-2)$
15	1	1	−1	0	$0\bar{0}\dot{+}(-1)$
16	0	0	−1	0	$0\bar{0}\dot{+}(-1)$
17	−1	−1	−2	0	$0\bar{0}\dot{+}(-2)$
18	0	−0	−2	−1	$-1\bar{0}\dot{+}(-2)$
19	−1	−1	−3	−1	$-1\bar{0}\dot{+}(-3)$
20	0	−0	−3	−2	$-2\bar{0}\dot{+}(-3)$

$$T_2 = \bigcup_{p\in\pi} p^2 \cdot \{1, T_1\}$$

Since this union is disjoint, we have

$$S(x) = \sum_{p\in\pi,\, p\leq\sqrt{x}} \left((-1)^{k(p^2)} + \sum_{l\in T_1,\, lp^2\leq x} (-1)^{k(lp^2)} \right)$$

$$S(x) = \sum_{p\in\pi,\, p\leq\sqrt{x}} \left(1 + \sum_{l\in T_1,\, l\leq\frac{x}{p^2}} (-1)^{k(l)} \right)$$

$$S(x) = \sum_{p\leq\sqrt{x}} M\left(\left[\frac{x}{p^2} \right] \right)$$

■

Example 11.1

$$S(16) = \sum_{p \le \sqrt{16}} M\left(\left[\frac{x}{p^2}\right]\right) = M[4] + M[1] = (-1) + (1) = 0$$

11.3 Connection to the Riemann Hypothesis

Theorem 11.3 Let $f\colon [0,\infty) \to \mathbb{R}$ be a function. Then $\lim_{x\to\infty} \frac{f(x)}{x^{\frac{1}{2}+\varepsilon}} = 0$ for every $\varepsilon > 0$ if and only if for every $\varepsilon > 0$ there is $C(\varepsilon)$ such that $\frac{|f(x)|}{x^{\varepsilon+\frac{1}{2}}} \le C(\varepsilon)$.

Proof 11.3 Let us assume that for every ε > 0 there is C(ε) such that

$$\frac{|f(x)|}{x^{\varepsilon+\frac{1}{2}}} \le C(\varepsilon)$$

$$\frac{|f(x)|}{x^{\varepsilon+\frac{1}{2}}} = \frac{|f(x)|}{x^{\frac{\varepsilon}{2}+\frac{1}{2}}} \cdot \frac{1}{x^{\frac{\varepsilon}{2}}} \le C\left(\frac{\varepsilon}{2}\right) \cdot \frac{1}{x^{\frac{\varepsilon}{2}}}$$

and therefore $\lim_{x\to\infty} \frac{f(x)}{x^{\frac{1}{2}+\varepsilon}} = 0$ The opposite direction is trivial.

■

Theorem 11.4 If for every $\varepsilon > 0$, $\lim_{x\to\infty} \frac{M(x)}{x^{\frac{1}{2}+\varepsilon}} = 0$, then for every $\varepsilon > 0$ $\lim_{x\to\infty} \frac{S(x)}{x^{\frac{1}{2}+\varepsilon}} = 0$.

Proof 11.4 Suppose that for every $\varepsilon > 0$, $\lim_{x\to\infty} \frac{M(x)}{x^{\frac{1}{2}+\varepsilon}} = 0$, then for every $\varepsilon > 0$ there is $C(\varepsilon)$

$$\frac{|M(x)|}{x^{\varepsilon+\frac{1}{2}}} \le C(\varepsilon)$$

$$|M(x)| \le C(\varepsilon) x^{\frac{1}{2}+\varepsilon}$$

$$S(x) = \sum_{p \le \sqrt{x},\, p\in\pi} M\left(\left[\frac{x}{p^2}\right]\right)$$

$$S(x) \le \sum_{p \le \sqrt{x},\, p\in\pi} C(\varepsilon)\left[\frac{x}{p^2}\right]^{\frac{1}{2}+\varepsilon} \le \sum_{p \le \sqrt{x},\, p\in\pi} C(\varepsilon)\left(\frac{x}{p^2}\right)^{\frac{1}{2}+\varepsilon}$$

$$S(x) \le \sum_{p \le \sqrt{x},\, p\in\pi} C(\varepsilon)\left[\frac{x}{p^2}\right]^{\frac{1}{2}+\varepsilon} \le C(\varepsilon) x^{\frac{1}{2}+\varepsilon} \sum_{p \le \sqrt{x},\, p\in\pi} \frac{1}{p^{1+2\varepsilon}}$$

$$\sum_{p \le \sqrt{x},\, p\in\pi} \frac{1}{p^{1+2\varepsilon}} \le \sum_{n=1}^{\infty} \frac{1}{n^{1+2\varepsilon}} = K(\varepsilon)$$

n	#1	#-1	#0	#-0	#+0	1%	-1%	0%	% 0-	0+	S(n)	M(n)	S(n)/n^0	M(n)/n^0.75
30000	9130	9112	2840	4527	4390	30.4333	30.3733	9.46667	15.09	14.6333	-137	18	-0.0601	0.0078964
50000	15212	15189	4739	7518	7341	30.424	30.378	9.478	15.036	14.682	-177	23	-0.0529	0.0068786
500000	151976	151982	47479	74589	73973	30.3952	30.3964	9.4958	14.9178	14.7946	-616	-6	-0.0328	-0.0003191
2000000	607815	608062	189990	297750	296382	30.3908	30.4031	9.4995	14.8875	14.8191	-1368	-247	-0.0257	-0.0046443
3000000	911940	911833	284988	446349	444889	30.398	30.3944	9.4996	14.8783	14.8296	-1460	107	-0.0203	0.0014844
5000000	1519462	1520171	475007	743687	741672	30.3892	30.4034	9.50014	14.8737	14.8334	-2015	-709	-0.0191	-0.0067053
10000000	3040164	3039127	950065	1486414	1484229	30.4016	30.3913	9.50065	14.8641	14.8423	-2185	1037	-0.0123	0.0058315
20000000	6078811	6079764	1900176	2972640	2968609	30.3941	30.3988	9.50088	14.8632	14.843	-4031	-953	-0.0135	-0.0031865

Fig. 11.1 Result of the computer simulation

$$S(x) \leq K(\varepsilon)\cdot C(\varepsilon)x^{\frac{1}{2}+\varepsilon}$$

$$\frac{S(x)}{x^{\frac{1}{2}+\varepsilon}} \leq K(\varepsilon)\cdot C(\varepsilon),$$

and therefore

$$\lim_{x\to\infty}\frac{S(x)}{x^{\frac{1}{2}+\varepsilon}} = 0$$

■

Theorem 11.5 The Riemann hypothesis is equivalent to the following statement: For every $\varepsilon > 0$, $\lim_{x\to\infty}\frac{\tilde{M}(x)}{x^{\frac{1}{2}+\varepsilon}} = 0$.

Proof 11.5 Since $x^{\frac{1}{2}+\varepsilon}$ is real, we have $\frac{\tilde{M}(x)}{x^{\frac{1}{2}+\varepsilon}} = \frac{S(x)}{x^{\frac{1}{2}+\varepsilon}}\overline{0}\dot{+}\frac{M(x)}{x^{\frac{1}{2}+\varepsilon}}$. Hence by definition, $\lim_{x\to\infty}\frac{\tilde{M}(x)}{x^{\frac{1}{2}+\varepsilon}} = 0$ if and only if $\lim_{x\to\infty}\frac{M(x)}{x^{\frac{1}{2}+\varepsilon}} = \lim_{x\to\infty}\frac{S(x)}{x^{\frac{1}{2}+\varepsilon}} = 0$. By Theorem 11.4, this is true for every $\varepsilon > 0$ if and only if $\lim_{x\to\infty}\frac{M(x)}{x^{\frac{1}{2}+\varepsilon}} = 0$ for every $\varepsilon > 0$. The latter statement is known to be equivalent to the Riemann hypothesis.

■

We present the result of simulation in Fig. 11.1

While $M(n)$ is moving from positive to negative as expected, $S(n)$ is negative and decreasing. Further investigation is necessary with a supercomputer to check if this can be evidence for a counterexample to the Riemann hypothesis. Another direction for further research is to prove that if for every $\varepsilon > 0$ $\lim_{x\to\infty}\frac{S(x)}{x^{\frac{1}{2}+\varepsilon}} = 0$, then for every $\varepsilon > 0$ $\lim_{x\to\infty}\frac{M(x)}{x^{\frac{1}{2}+\varepsilon}} = 0$.

Reference

Edwards, H.M.: Riemann's Zeta Function, vol. 58. Courier Corporation (2001)

Chapter 12
Soft Probability and Entropy

This chapter extends classical probability theory using Soft logic. To do so, we start from a fundamental distinction in continuous probability theory, according to which the probability of a specific event occurring is zero. This fact is of course illogical. How can an event that actually occurred have an a priori probability of zero? (Fivel, Klein, and Maimon, 2021, 2023)

12.1 Introduction to Soft Probability

Using Soft numbers, we propose a new mathematical model in which the probability of a specific event occurring is not an absolute zero but a Soft zero. Using this distinction, we define Soft probability and discuss complementary, disjoint, and intersecting events, as well as Soft conditional probability. Using Soft probability, we then define the concepts of expectation, variance, and entropy. We conclude the chapter with the possibility of applying Soft probability to developing decision trees using Soft numbers.

Classical probability theory was developed by Fermat and Pascal, who analyzed gambling games that end early and in which the pot needs to be divided between the two players. The question they solved was how to fairly divide the pot according to the chances of each of the two players winning the game. The axiomatic foundation of probability theory occurred only in the 1930s, by Kolmogorov. Notably, in formulating the axioms of probability theory, Kolmogorov did not refer to the great debate of the time between physicists regarding the probabilistic interpretation of quantum mechanics.

Soft logic develops a new concept of modeling and dealing with uncertainty in time and space, which refers not only to the future but also to the present, that is, how we interpret reality. For this purpose, we have developed a new mathematical language that bridges between objective and subjective reality. Soft probability theory was developed to model processes and phenomena involving randomness

M. Klein, O. Maimon, *Foundations of Soft Logic*,
https://doi.org/10.1007/978-3-031-58233-2_12

and uncertainty that refer also to the present and to the different possible interpretations of reality by the observer.

The probability of a continuous random variable is defined using the probability density function. We use the density function of the distribution to approximate the probability of a continuous random variable X being close to x as follows:

$$\Pr(x < X \leq x + \Delta x) \approx f_X(x)\Delta x,$$

where $\Delta x > 0$ is an infinitesimal value describing the accuracy level of the probability.

Continuous random variables have two properties that raise the need to expand them using Soft numbers, as follows:

(1) There is no distinction at all between strict and non-strict inequality, meaning $\Pr(X \leq x) = \Pr(X < x)$.
(2) Equality collapses to zero, meaning $\Pr(X = x) = 0$.

Although any value of $x \in S_x$ (S_x denotes the support of X) is possible, the probability that X will be equal to any specific value $x \in S_x$ is (almost surely) zero.

In this way we actually lose information regarding the probability of a continuous random variable taking on a specific value. The event $X = x$ is possible, but the probability of it occurring is zero. This seems like a paradoxical situation.

We can express the zero probability of the event "$X = x$" through the fact that as we get closer by Δx to X, meaning $\Pr(X = x) = f_X(x) \cdot 0$. This equation presents the probability $\Pr(X = x)$ as the product of zero by the density function factor $f_X(x)$for any x. Instead of assuming that $\Pr(X = x)$ is an absolute zero, we can assign it the product of a Soft zero of $f_X(x)$ and compare different probability values for different observation values x. This approach can be implemented by using Soft numbers, with which we can give a probabilistic interpretation to a continuous random variable that distinguishes between strict and non-strict inequality in the probability function.

12.2 Defining Soft Probability

We will use a different notation to define the cumulative distribution function of a continuous random variable in a different way: $\mathrm{P_s}(X \leq x) = F_x(1 \cdot \bar{0} \dotplus x)$, where Ps $(\cdot)$ is a proposed type of probability function, denoted as "Soft probability" (instead of the usual probability notation $\Pr(\cdot)$ or $\mathrm{P}(\cdot)$, and $F_x(\cdot)$ is the usual cumulative distribution function of the random variable X, but it is applied to the Soft number $1 \cdot \bar{0} \dotplus x$.

Our motivation is to create an alternative estimate of the probability on the left side of the equation, so that we can distinguish between $\mathrm{P_s}(X < x)$ and $\mathrm{P_s}(X \leq x)$ for a continuous random variable X (meaning, $\mathrm{P_s}(X < x) \neq \mathrm{P_s}(X \leq x)$).

It appears that estimating the Soft number on the right side of the cumulative distribution function creates this distinction.

This can be decomposed as follows: $F_X(1 \cdot \bar{0} \dot{+} x) \stackrel{\text{def}}{=} f_X(x)\bar{0} \dot{+} F_X(x)$. We can decompose the left side by splitting the event "X ≤ x" to the disjoint union "$X = x \dot{\cup} X < x$."

In regular probability, we have the following well-known identities:

$$\Pr(X \le x) = \overbrace{\Pr(X = x)}^{=0} + \Pr(X < x) = \Pr(X < x)$$

Therefore we do not have a distinction between $\Pr(X \le x)$ and $\Pr(X < x)$.

We distinguish between $P_s(X \le x)$ and $P_s(X < x)$ through the following definition for $P_s(X \le x)$:

$$P_s(X \le x) \stackrel{\text{def}}{=} P_S(X = x) + P_S(X < x)$$

such that we define the left-hand terms as

$$P_S(X = x) \stackrel{\text{def}}{=} f_X(x)\bar{0}$$

$$P_S(X < x) \stackrel{\text{def}}{=} F_X(x) \equiv \Pr(X < x)$$

Through this definition we achieve a distinction between $P_s(X \le x)$ and $P_S(X < x)$ and also provide an interpretation of $P_S(X = x)$ as having an infinitesimally small but nonzero value due to the $\bar{0}$ factor of the density function.

Example 12.1 Gaussian distribution

Let X be a Gaussian random variable parameterized by a mean μ and a variance σ^2 (denoted $X \sim N(\mu, \sigma^2)$). The PDF of X is well known as

$$f_X\left(x, \mu, \sigma^2\right) = \frac{1}{\sqrt{2\pi\sigma^2}} e^{-\frac{1}{2\sigma^2}(x-\mu)^2}$$

The maximum of the PDF, $\max_x f_X(x, \mu, \sigma^2) = \frac{1}{\sqrt{2\pi\sigma^2}}$, occurs at $x = \mu$. We would like to have a higher probability, as X is closer to μ, e.g., $P_s(X = \mu) > P_s(X = x), \forall\, x \neq \mu$.

We have the following definition of Soft probability in the Gaussian case:

$$P_s(X = x) = \frac{1}{\sqrt{2\pi\sigma^2}} e^{-\frac{1}{2\sigma^2}(x-\mu)^2}\bar{0},$$

which presents an absolute low probability of X having an exact value x but a relatively high probability when X is closer to μ.

12.3 Complements, Unions, Intersections, and Soft Conditional Probability

In this subsection, we extend the concept of "Soft probability" to complements, unions, and intersections of events and to conditional probability. In the first part, we show the results for complements, unions, and intersections referring to events with zero probability in the classical probability theory sense. In the second part, we show the results for Soft conditional probability, based on Kolmogorov's definition and Bayes' theorem.

12.3.1 Complements, Unions, and Intersections

As is known, the probability of the complementary event A^C is given by $\Pr(A^C) = 1 - \Pr(A)$. The Soft probability of a complement is defined similarly: $\mathrm{Ps}(A^C) = 1 - \mathrm{Ps}(A)$. Therefore, we have the following complementary probability for a continuous random variable X:

$$\mathrm{Ps}(X \neq x) = 1 - \mathrm{Ps}(X = x) = [-f_X(x)]\bar{0} + 1$$

This equation distinguishes between different values of x in the case of the event $X \neq x$ almost surely having a probability of 1 because of the Soft zero term $[-f_X(x)]\bar{0}$. This is analogous to the event $X \neq x$ having a probability of zero, which is almost surely true because of the Soft zero term $[-f_X(x)]\bar{0}$.

In order to analyze unions and intersections of events, we will distinguish in principle between two cases as follows: unions and intersections between individual events of the type $X = x$, $X = y$, etc., and between a single event of the type $X = x$ and a range event of the type $a \leq X \leq b$.

For any $x \neq y$, we have that the events $X = x$ and $X = y$ are disjoint, and for a union we have

$$\mathrm{Ps}(\mathrm{X} = \mathrm{x} \cup \mathrm{X} = \mathrm{y}) = \mathrm{Ps}(\mathrm{X} = \mathrm{x}) + \mathrm{Ps}(\mathrm{X} = \mathrm{y}) = [f_X(\mathrm{x}) + f_X(\mathrm{y})]\bar{0}$$

For an intersection of two events, we have $\mathrm{Ps}(\mathrm{X} = \mathrm{x} \cap \mathrm{X} = \mathrm{y}) = 1_{x=y} f(\mathrm{x})\bar{0}$, where the indicator $1_{x=y}$ is zero when $x \neq y$. More generally, for the set $\{x_i\}_{i=1}^{n}$ with distinct values, we have the Soft probabilities

$$Ps\left(\bigcup_{i=1}^{n} X = x_i\right) = \sum_{i=1}^{n} Ps(X = x_i) = \left[\sum_{i=1}^{n} f_X(x_i)\right]\bar{0}$$

and

$$Ps\left(\bigcap_{i=1}^{n} X = x_i\right) = 1_{\forall i,j \in \{1,2,\ldots,n\}\ x_i = x_j}\ f_X(x_i)\bar{0}$$

In order to analyze unions and intersections between a single event of the type $X = x$ and a range event of the type $a \leq X \leq b$, we need to distinguish between cases where x is or is not between a and b. In addition, we need to distinguish between a case of strict inequality $a < X < b$ and non-strict inequality $a \leq X \leq b$. For simplicity, assume $a < b$, and without loss of generality (WLOG), assume $x \neq a$ and $x \neq b$.

For the case of strict inequality $a < X < b$, we have the union

$$\mathrm{Ps}(X = x \cup a < X < b) = \left[1_{x \notin (a,b)} f_X(x) + f_X(a) + f_X(b)\right]\bar{0} \dot{+} [F_X(b) - F_X(a)]$$

and the intersection

$$\mathrm{Ps}(X = x \cap a < X < b) = 1_{x \in (a,b)} f_X(x)\bar{0}.$$

This is a Soft number when x is in the interval (a, b) and a real number when it is not in the interval. This is a Soft zero when x is in (a, b) and an absolute zero when it is not in the interval.

For the case of non-strict inequality $a \leq X \leq b$, we have the union

$$\mathrm{Ps}(X = x \cup a \leq X \leq b) = [1_{x \notin [a,b]} f_X(x) + f_X(a) + f_X(b)]\bar{0} \dot{+} [F_X(b) - F_X(a)]$$

and the intersection

$$\mathrm{Ps}(X = x \cap a \leq X \leq b) = \left[1_{x \in [a,b]} f_X(x)\right]\bar{0}.$$

The term $f_X(b)$ is added to the Soft zero part.

According to De Morgan's law, there is a relationship between the union and intersection of two events A,B: Pr $(A \cup B) =$ Pr $(A) +$ Pr $(B) -$ Pr $(A \cap B)$.

De Morgan's law also holds for Soft probabilities. For example, in the case of

$$A = \{X = x\}, B = \{a \leq X \leq b\} \text{ and } x \notin [a, b],$$

we have

$$\mathrm{Ps}(X = x \cup a \leq X \leq b) = \mathrm{Ps}(X = x) + \mathrm{Pr}(a \leq X \leq b) - Pr(X = x \cap a \leq X \leq b)$$

The left-hand side is

$$[f_X(x) + f_X(a) + f_X(b)]\bar{0} \dot{+} [F_X(b) - F_X(a)]$$

and the right-hand side is

$$f_X(x)\bar{0} + [(f_X(a) + f_X(b))\bar{0} \dot{+} \{F_X(b) - F_X(a)\}] - 0$$

so the left-hand side equals the right-hand side, and therefore we have a "Soft De Morgan's law":

$$\mathrm{Ps}(A \cup B) = \mathrm{Ps}(A) + \mathrm{Ps}(B) - \mathrm{Ps}(A \cap B)$$

In the next subsection, we show the results for Soft conditional probability, with reference to Kolmogorov's definition and Bayes' theorem.

12.3.1.1 Conditional Probability

Recall Kolmogorov's definition of conditional probability $\Pr(A|B) = \frac{\Pr(A\cap B)}{\Pr(B)}$ and Bayes' theorem $\Pr(A|B) = \frac{\Pr(B|A)\Pr(A)}{\Pr(B)}$. We define "Soft conditional probability" similarly. For example, for $x, y \in S_X$, let $A = \{X = x\}$, $B = \{X = y\}$, and on the right-hand side of Kolmogorov's definition we have

$$\frac{Ps(X = x \cap X = y)}{Ps(X = y)} = \frac{1_{x=y} f_X(y)\bar{0}}{f_X(y)\bar{0}} = \frac{1_{x=y}\bar{0}}{1 \cdot \bar{0}}$$

If we define $\frac{1\cdot\bar{0}}{1\cdot\bar{0}} = 1$ and $\frac{0\cdot\bar{0}}{1\cdot\bar{0}} = 0$, the Soft conditional probability is given by

$$\mathrm{Ps}(X = x | X = y) = 1_{x=y}$$

In this case, we have a trivial equality with the real values 0 or 1. For comparison with Bayes' theorem:

$$\frac{\mathrm{Ps}(X = y | X = x) Ps(X = x)}{Ps(X = y)} = \frac{1_{y=x} f_X(x)\bar{0}}{f_X(y)\bar{0}} = 1_{x=y}$$

Now consider the more general case of $x \in S_X$.

Let $A = \{X = x\}$, $B = \{a \leq X \leq b\}$, with $x, a, b \in S_X$ such that $a < b$, $x \neq a$ and $x \neq b$.

On the right-hand side of Kolmogorov's definition, we have

$$\frac{\mathrm{Ps}(X = x \cap a \leq X \leq b)}{Ps(a \leq X \leq b)} = \frac{1_{x\in[a,b]} f_X(x)\bar{0}}{[f_X(a) + f_X(b)]\bar{0} \dot{+} [F_X(b) - F_X(a)]}$$

In applying Bayes' theorem, we have

$$\frac{Ps(a \le X \le b | X = x) Ps(X = x)}{Ps(a \le X \le b)} = \frac{[\mathrm{Ps}(a \le x \le b | X = x)] f_X(x) \bar{0}}{[f_X(a) + f_X(b)] \bar{0} \dotplus [F_X(b) - F_X(a)]},$$

where

$$\mathrm{Ps}(a \le x \le b | X = x) = \mathrm{Ps}(a \le x \le b) = 1_{x \in [a,b]}$$

Both formulas—Kolmogorov's definition and Bayes' theorem—are equal, and therefore

$$Ps(X = x | a \le X \le b) = \frac{1_{x \in [a,b]} f_X(x) \bar{0}}{[f_X(a) + f_X(b)] \bar{0} \dotplus [F_X(b) - F_X(a)]}.$$

The right-hand side can be simplified using the property

$$\frac{A\bar{0}}{B \dotplus C\,\check{0}} = \frac{A\bar{0}}{B \dotplus C\,\check{0}} \cdot \frac{B \dotplus (-C)\,\check{0}}{B \dotplus (-C)\,\check{0}} = \frac{AB\bar{0}}{B^2} = \frac{A\bar{0}}{B}$$

so that we get the following Soft conditional probability for a given non-strict inequality condition:

$$\mathrm{Ps}(X = x | a \le X \le b) = \frac{1_{x \in [a,b]} f_X(x) \bar{0}}{F_X(b) - F_X(a)}$$

For a given strict inequality condition we have

$$\mathrm{Ps}(X = x | a < X < b) = \frac{1_{x \in (a,b)} f_X(x) \bar{0}}{F_X(b) - F_X(a)}$$

12.4 Soft Expectation, Variance, and Entropy

In this part, we define Soft expectation, Soft variance, and Soft entropy. First, we focus on the definitions of expectation and variance, with a reminder of their original well-known definitions, and then generalize them to Soft numbers. We then do the same for entropy.

12.4.1 Soft Expectation and Variance

Recall that for the expectation definition of a random variable X with support S_x

$$E(X) = \int_{S_x} x dF_X(x) = \mu_{X,}$$

where $E(\cdot)$ is the expectation operator defined by the above integral, and we denote its result by μ_X (sometimes called the mean). For a continuous random variable, the expectation is defined by $E(X) = \int_{S_x} x f_X(x) dx$ and for a discrete random variable $E(X) = \sum\limits_{x \in S_X} x \Pr(X = x)$.

The variance of a random variable is the expectation of the squared error from its mean, that is

$$\mathrm{Var}(X) = E\left[(X - \mu_X)^2\right] = \int_{S_X} (x - \mu_X)^2 \, dF_X(x) = \sigma_X^2,$$

$$\mathrm{Var}(X) = \int_{S_X} (x - \mu_X)^2 f_X(x) dx$$

for the continuous case, and

$$\mathrm{Var}(X) = \sum_{x \in S_X} (x - \mu_X)^2 Pr(X = x)$$

for the discrete case.

Assume X is a continuous random variable and $\{x_i\}_{i=1}^n$ and $\{(a_j, b_j)\}_{j=1}^m$ are a set of numbers and a set of disjoint intervals in the support S_x. Also assume $\{x_i\}_{i=1}^n$ and $\{(a_j, b_j)\}_{j=1}^m$ are disjoint. Without loss of generality (WLOG), we refer to open intervals; otherwise, we can take out the interval endpoints aj, bj from the interval a_j, b_j and include them in the set of numbers $\{x_i\}_{i=1}^n$. Under the above assumption, we define the Soft expectation of X conditional on it being in the union of $\{x_i\}_{i=1}^n$ and $\{(a_j, b_j)\}_{j=1}^m$, that is

$$\begin{aligned} \mathrm{Es}\left(X | X \in \{x_i\}_{i=1}^n \cup \{(a_j, b_j)\}_{j=1}^m\right) &= \sum_{i=1}^n x_i Ps(X = x_i) \dot{+} \sum_{j=1}^m \int_{a_j}^{b_j} x f_x(x) dx \\ &= \sum_{i=1}^n x_i f_X(x_i) \cdot \bar{0} \dot{+} \sum_{j=1}^m \int_{a_j}^{b_j} x f_x(x) dx = \nu_X \bar{0} \dot{+} \kappa_X, \end{aligned}$$

where $\mathrm{Es}(\cdot)$ is a new notation for the Soft expectation operator. Here we are using the concept of conditional expectation; however, instead of computing the expectation of a given random variable conditioned on another random variable, the condition is given on the same variable but within a certain set of discrete points and intervals.

Note that X is a random variable with real values, and also all the discrete points $\{x_i\}_{i=1}^n$ and all the intervals $\{(a_j, b_j)\}_{j=1}^m$ are real. However, due to the Soft probability terms $Ps(X = x_i)$, the result on the right-hand side is a Soft number.

For simplicity, we denote the real part of the Soft expectation by κ_X and the Soft part by ν_X. With this definition, the Soft part ν_X adds new information about the mean of the continuous random variable X being at discrete points $\{x_i\}_{i=1}^n$.

This value would be zero without this Soft expectation definition. We can define the Soft expectation of a function $g(X)$ mapping from real numbers to real or Soft numbers as

$$Es(g(X)|X \in \{x_i\}_{i=1}^n \cup \{(a_j, b_j)\}_{j=1}^m)$$
$$= \sum_{i=1}^n g(x_i)\ Ps(X = x_i) \dot{+} \sum_{j=1}^m \int_{a_j}^{b_j} g(x) f_X(x) dx = \sum_{i=1}^n g(x_i) f_X(x_i)$$
$$\cdot \bar{0} \dot{+} \sum_{i=1}^m \int_{a_j}^{b_j} g(x) f_X(x) dx.$$

With this concept, we define the Soft variance (denoted by Vs) of X conditional on it being in the union of $\{x_i\}_{i=1}^n$ *and* $\{(a_j, b_j)\}_{j=1}^m$.

Using the zeros property of $\bar{0}$ (Axiom 3) and the derivative property, we have

$$Vs(X|X \in \{x_i\}_{i=1}^n \cup \{(a_j,b_j)\}_{j=1}^m) = Es((X - (\nu_X \bar{0} \dot{+} \kappa_X))^2) | X$$
$$\in \{x_i\}_{i=1}^n \cup \{(a_j,b_j)\}_{j=1}^m) = \sum\nolimits_{i=1}^n [\nu_X \bar{0} \dot{+} (\kappa_X - x_i)]^2 f_X(x_i) \bar{0} \dot{+} \sum_{j=1}^m \int_{a_j}^{b_j} (\nu_X \bar{0} \dot{+} \kappa_X - x)^2 f_X(x) dx$$
$$= \left[\sum_{i=1}^n (\kappa_X - x_i)^2 f_X(x_i) + 2\nu_X \sum_{j=1}^m \int_{a_j}^{b_j} (\kappa_X - x) f_X(x) dx\right] \bar{0} \dot{+} \sum\nolimits_{j=1}^m \int_{a_j}^{b_j} (\kappa_X - x)^2 f_X(x) dx.$$

We want to simplify this last equation, especially the Soft part. Let's denote:

$$\gamma_{1\ X}^2 = \sum_{i=1}^n (\kappa_X - x_i)^2 f_X(x_i) \geq 0$$

$$\gamma_{2\ X} = \sum_{j=1}^m \int_{a_j}^{b_j} (\kappa_X - x) f_X(x) dx$$

$$\lambda_X^2 = \sum_{j=1}^{m} \int_{a_j}^{b_j} (\kappa_X - x)^2 f_X(x)dx \geq 0$$

Using the linearity of the integral, we can simplify $\gamma_{2\ x}$ as

$$\gamma_{2\ x} = \kappa_X \sum_{j=1}^{m} \int_{a_j}^{b_j} f_X(x)dx - \sum_{j=1}^{m} \int_{a_j}^{b_j} xf_X(x)dx = \kappa_X \sum_{j=1}^{m} \left[F_X(b_j) - F_X(a_j)\right] - \kappa_X = - \kappa_X \left[1 - \sum_{j=1}^{m} \left[F_X(b_j) - F_X(a_j)\right]\right]$$

We can see that $1 - \sum_{j=1}^{m} \left[F_X(b_j) - F_X(a_j)\right] > 0$.

Now we simplify the definition of Soft variance:

$$Vs\left(X|X \in \{x_i\}_{i=1}^{n} \cup \left\{(a_j, b_j)\right\}_{j=1}^{m}\right) = \left[\sum_{i=1}^{n} (\kappa_X - x_i)^2 f_X(x_i) - 2\nu_X \kappa_X \left[1 - \sum_{j=1}^{m} \left[F_X(b_j) - F_X(a_j)\right]\right]\right]\bar{0} + \sum_{j=1}^{m} \int_{a_j}^{b_j} (\kappa_X - x)^2 f_X(x)dx = \left[\gamma_{1\ X}^2 - 2\nu_X \gamma_{2\ x}\right]\bar{0} + \lambda_X^2$$

The real part λ_X^2 is nonnegative (equal to zero if $x \equiv \kappa_X, \forall\, x \in (a_j, b_j), j = 1, 2\ldots,$ m, meaning X is deterministic), which makes sense in terms of the original variance definition. However, in the Soft part $\gamma_{1\ X}^2 - 2\nu_X \gamma_{2\ x}$, there is an interesting phenomenon. On one hand, we have a nonnegative term $\gamma_{1\ X}^2$ (equal to zero if $x_i \equiv \kappa_X, \forall\, i = 1, 2\ldots n$). On the other hand, the sign of the term $-2\nu_X \gamma_{2\ x}$ combined with the Soft part depends on the sign of ν_X (which depends on the sign of κ_X), so potentially the Soft part can be negative. The Soft part adds additional information about the variance of the random variable. We will examine applications of Soft variances with negative Soft parts.

In the next subsection, we continue to the definition of Soft entropy, inspired by the concepts in the previous subsections.

12.4.2 Soft Entropy

Recall that the entropy of a discrete random variable X with support S_x and probability mass function p_X is defined by

$$H(X) = -\mathrm{E}(\log p_X(X)) = -\sum_{x \in SX} p_X(x) \log p_X(x).$$

For a continuous case (usually referred to as differential entropy) of a continuous random variable $X \sim f_X$, we have

$$H(X) = -E(\log f_X(X)) = -\int_{S_x} f_X(x) \log f_X(x) dx,$$

where (in both definitions) the logarithm base can be chosen as a suitable positive real number, etc., depending on the application. In this work, there is no emphasis on a specific base but as before, we refer to a continuous random variable X conditional on it being in the union of $\{x_i\}_{i=1}^n$ *and* $\{(a_j, b_j)\}_{j=1}^m$. With these definitions, we get the following definition for Soft entropy Hs(·):

$$\begin{aligned}
&\mathrm{Hs}(\mathrm{X}|\mathrm{X} \in \{x_i\}_{i=1}^n \cup \{(a_j, b_j)\}_{j=1}^m) \\
&= \sum_{i=1}^n -\mathrm{Ps}(\mathrm{X}=\mathrm{x}_i) \log(\mathrm{Ps}(\mathrm{X}=\mathrm{x}_i)) \dot{+} \sum_{j=1}^m - \int_{a_j}^{b_j} \mathrm{f_X}(\mathrm{x}) \log(\mathrm{f_X}(\mathrm{x})) \mathrm{dx} \\
&= \sum_{i=1}^n [-\mathrm{f_X}(\mathrm{x}_i) \cdot \bar{0}][\log(\mathrm{f_X}(\mathrm{x}_i) \cdot \bar{0}))] \dot{+} \sum_{j=1}^m - \int_{a_j}^{b_j} \mathrm{f_X}(\mathrm{x}) \log(\mathrm{f_X}(\mathrm{x})) \mathrm{dx} \\
&= \left[\sum_{i=1}^n [-\mathrm{f_X}(\mathrm{x}_i)][\bar{0} \log(\bar{0})] \dot{+} [\sum_{i=1}^n -f_X(x_i) \log(f_X(x_i)) \right] \bar{0} \\
&\quad + \sum_{j=1}^m - \int_{a_j}^{b_j} \mathrm{f_X}(\mathrm{x}) \log(\mathrm{f_X}(\mathrm{x})) \mathrm{dx} = \mathrm{h}_1 \cdot [\bar{0} \log(\bar{0})] \dot{+} \mathrm{h2} \cdot \bar{0} \dot{+} \mathrm{h3} \cdot 1,
\end{aligned}$$

where $\mathrm{h2} \cdot \bar{0} \dot{+} \mathrm{h3} \cdot 1 = -\mathrm{Es}(\log(\mathrm{f_X}(\mathrm{X})|\mathrm{X} \in \{x_i\}_{i=1}^n \cup \{(a_j, b_j)\}_{j=1}^m)$ with $\mathrm{g(x)} = \log(\mathrm{f_X}(\mathrm{x}))$ and $h_1 = \sum_{i=1}^n (-f_X(x_i))$.

The Soft entropy of X is a linear combination of the objects $\bar{0}, \bar{0} \log(\bar{0})$, and 1. The question is how to evaluate the object $\bar{0} \log(\bar{0})$. One option would be an absolute zero, meaning $\bar{0} \log(\bar{0}) = 0$.

Observation:

$$\lim_{x\to 0^+} x^x = 1 \Rightarrow \lim_{x\to 0^+} \mathrm{xlog(x)} = 0$$

$$e^x = \sum_{n=0}^{\infty} \frac{x^n}{n!} = \frac{x^0}{0!} + \frac{x^1}{1!} + \frac{x^2}{2!} + \frac{x^3}{3!} + \dots$$

$$1 = e^0 = \sum_{n=0}^{\infty} \frac{0^n}{n!} = \frac{0^0}{0!} + \frac{0^1}{1!} + \frac{0^2}{2!} + \frac{0^3}{3!} + \dots = \frac{0^0}{0!} + (0) = 0^0$$

Another option would be to define it as a new type of "Soft zero" object, for example, a new axis which is a continuum of $\bar{\bar{0}} = \bar{0}\,\log(\bar{0})$, with a zeros property of $\bar{\bar{0}}^2 = 0$. With the second option, we have additional information about the entropy of X, not only through the zero axis but also through a potential new axis, $\bar{\bar{0}} = \bar{0}\log(\bar{0})$.

12.5 Decision Trees Based on Soft Mutual Information

Decision trees are simple yet successful techniques for predicting and explaining the relationship between some measurements of an item and its target value (Maimon and Rokach 2014). In most decision tree inducers, discrete splitting functions (also known as splitting criteria) are univariate, i.e., an internal node is split according to the value of a single attribute. Consequently, the inducer searches for the best attribute upon which to perform the split. A splitting criterion of a random variable X (representing the features) and a random variable Y (representing the labels) has the following structure: splitting criteria $(Y; X) = C(Y) - C(Y \mid X)$, where $C(\cdot)$ is an expectation of some cost function. In the case when the cost function is the entropy [i.e., $C(\cdot) = H(\cdot)$], we refer to the splitting criteria as an information gain that is mutual information between X and Y, denoted by $I(Y;X) = H(Y) - H(Y \mid X) = H(Y) + H(X) - H(Y,X)$, which also can be written as a KLD between the joint PDF $f_{X,\,Y}$ and the PDF product $f_X f_Y$, i.e., $I(Y;X) = D(f_{X,\,Y} \| f_X f_Y)$. In this section, we present mutual information (an example of a splitting criteria) as a Soft number, based on a joint PDF (and its related marginal PDFs) of two random continuous variables, but with a dataset consisting of single values and an interval. From here, the decision algorithm is clear. Suppose that X and Y are continuous random variables, such that X is within the union of $\{x_i\}_{i=1}^n$ *and* $\{(a_j, b_j)\}_{j=1}^m$, and Y is within the union of $\{y_i\}_{i=1}^N$ and $\{(A_j, B_j)\}_{j=1}^m$ (recall that the single points, i.e., $\{x_i\}_{i=1}^n$, and the intervals $\{(a_j, b_j)\}_{j=1}^m$ are disjoint). For simplicity, denote the following sets:

$$\mathcal{X} = \{x_i\}_{i=1}^{n} \cup \{(a_j, b_j)\}_{j=1}^{m}$$

$$\mathcal{Y} = \{y_i\}_{i=1}^{N} \cup \{(A_j, B_j)\}_{j=1}^{m}.$$

We can define the Soft mutual information Is(·) by the following equation after re-indexing:

$$\begin{aligned}
\text{Is}(Y;X|Y \in \mathcal{Y}, X \in \mathcal{X}) &= \text{Ds}(f_{X,Y} \| f_X f_Y | Y \in \mathcal{Y}, X \in \mathcal{X}) \\
&= \left[\sum_{j=1}^{N} \sum_{i=i}^{n} f_{X,Y}(x_i, y_j) \log\left(\frac{f_{X,Y}(x_i, y_j)}{f_X(x_i) f_Y(y_j)}\right)\right] \cdot \bar{0} \\
&\dot{+} \sum_{j=1}^{M} \sum_{i=1}^{m} \int_{A_j}^{B_j} \int_{a_i}^{b_i} f_{X,Y}(x, y) \log\left(\frac{f_{X,Y}(x, y)}{f_X(x) f_Y(y)}\right) dx dy,
\end{aligned}$$

Thus, we have an example of a Soft number as a splitting criteria that can be used in decision tree algorithms in the case of a data set consisting of single values and intervals. The definition of Soft mutual information is symmetric in X and Y (due to $I(X;Y) = I(Y;X)$ in the regular sense). We can present this in a less symmetric form, using the Bayes' law identity $f_{X,Y} = f_{Y|X} f_X$, so that we have

$$Is(Y;X|Y \in \mathcal{Y}, X \in \mathcal{X}) = \left[\sum_{j=1}^{N} \sum_{i=i}^{n} f_{Y|X}(y_j|x_i) f_X(x_i) \log\left(\frac{f_{Y|X}(y_j|x_i)}{f_Y(y_j)}\right)\right]$$

$$\cdot \bar{0} \dot{+} \sum_{j=1}^{M} \sum_{i=1}^{m} \int_{A_j}^{B_j} \int_{a_i}^{b_i} f_{Y|X}(y|x) f_X(x) \log\left(\frac{fY \mid X(y|x)}{f_Y(y)}\right) dx dy.$$

This representation is applicable, for example, for emphasizing X as an input and Y as an output to some channel.

References

Fivel, O., Klein, M., Maimon, O.: Soft decision trees. In: Rokach, L., Maimon, O., Shmueli, E. (eds.) Machine Learning for Data Science Handbook, pp. 143–170. Springer (2023)

Fivel, O., Klein, M., Maimon, O.: Decision trees with Soft numbers. Int. J. Circuits Syst. Signal Process. **15**, 1803–1816 (2021). https://doi.org/10.46300/9106.2021.15.194

Maimon, O., Rokach, L.: Data Mining with Decision Trees: Theory and Applications, 2nd edn. World Scientific, Singapore (2014). https://doi.org/10.1142/909

Chapter 13
Soft Logic and the Privacy Paradox

This chapter develops the idea of applying Soft logic to understanding aspects of human behavior. The psychologist Freud developed the theory of the unconscious: A person has inner drives and desires that he is not aware of, which nevertheless influence his behavior and feelings. This theory in itself suggests an inner component and an outer component that is revealed to a person. We can translate this fact using the terms of Soft logic. One real axis expresses the revealed world, and the zero axis expresses the inner and unconscious world.

13.1 Introduction to the Privacy Paradox

The ability to distinguish between reality and imagination is one of the important issues in a person's mental health. The psychologist Lacan used mathematical images taken from knot theory to describe mental states, including psychosis. He used the model of a Möbius strip to help a patient understand his mental state better and thereby assist him in coping with his mental problems. The psychologist Bion used mathematical images to describe a person's mental state. He described a state of elevation, which he called state O, that allows a person to live a good life despite the trauma he experienced in childhood. On a high level of awareness, the patient recognizes the problems he has and through this learns to adapt and live with the various psychological situations he is dealing with. Bion used various mathematical functions and understood that mathematical language can bring a person to a deeper understanding and acceptance of his mental state.

Bion and others raise the question of whether the subject of mental disorders can be better understood through mathematical language and what is the mathematical model that could point to a possible healing ability of the person. Understanding the connection between imagination and reality as a healing and mentally strengthening factor for individuals may be one of the applications of Soft logic and Soft numbers.

M. Klein, O. Maimon, *Foundations of Soft Logic*,
https://doi.org/10.1007/978-3-031-58233-2_13

We believe there is a split between the inner forces that drive a person and the things that are revealed and external and that we can express these things through Soft mathematics using Soft numbers. By mathematically describing this difference, we will be able to gain a deeper understanding of a person's mental state, and we will be able to develop practical tools for helping people understand and strengthen themselves mentally.

To illustrate the idea in a specific, way we demonstrate a problem called the privacy paradox. (Acquisti et al. 2015; Anic et al. 2019) This problem is related to people's attitude toward preserving their privacy in the cyber and internet world. The question is to what extent is a person willing to disclose personal details such as name, identity card number, address, credit card number, phone, and email address when required to do so on the Internet, for example, when one buys something or connects to a network group. It turns out that there is a big gap between the way a person defines himself and thinks and the way he acts in practice. It may be that a person says that he wishes to preserve his privacy, but in practice, when he is in front of the computer, he will disclose information that he said that he would not disclose. This gap is called the privacy paradox, and it has been extensively studied by researchers (Arpetti and Delmastro 2021; Baek 2014; Cover and Thomas 2006).

In the study we conducted, we tried to model the privacy paradox using Soft logic. We created a mathematical model that describes the privacy paradox. To do this, we also investigated people's technological literacy. We defined a mathematical formula that describes the privacy paradox in a Soft way using a Soft number. We defined two models to describe this paradox in the form of a Soft number.

We conducted an empirical test with a group of 132 people who filled out questionnaires to learn how they preserve privacy and how they actually behave. We also examined their level of technological literacy. We analyzed the research results using tables and diagrams. We presented the same data using the Soft axis system and obtained a different, yet revealing, picture of the same data. We saw that there is an order and structure that we did not notice in the standard axis system.

Note that this is only a preliminary study, but we managed to show how to generate an internal representation of a person using Soft numbers. The representation in the Soft axis system makes it possible to better see and understand the situation of paradox. It is possible to continue and apply this idea to a variety of different subjects in order to develop a psychological model that relies on and uses Soft logic and Soft numbers

13.2 Representing the Privacy Paradox Using Soft Logic

The privacy paradox refers to the gap between the way people define their level of privacy preservation on the Internet and their actual behavior. The digital era presents a significant issue regarding the way individuals preserve their privacy. Each person has two autonomous characteristics: *privacy concern*, indicating the level of

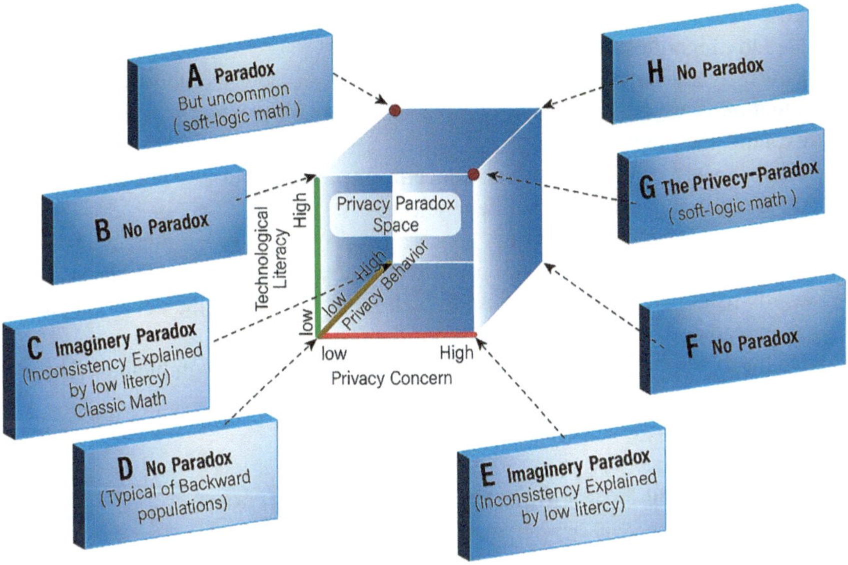

Fig. 13.1 Three-dimensional model of the privacy paradox

anxiety of that person to preserve their privacy, and *privacy behavior*, relating to actions taken by the individual to preserve their privacy. The significant gap that exists between these two characteristics is called the privacy paradox.

We leverage Soft logic to represent the privacy paradox differently (Hirschprung et al. 2022). Soft numbers expand the crisp zero into an infinite one-dimensional axis, thereby enabling the representation of contradictory states that exist simultaneously, i.e., a paradox. We develop a mathematical model for representing the privacy paradox using Soft numbers and demonstrate its application empirically. This new theory has the potential to deal with areas combining the Soft human reality and the solid technological reality.

The space spanned by the factors comprising the privacy paradox can be described by a three-dimensional model, as depicted in the following Fig. 13.1.

To explain the privacy paradox space, we identify the cube vertices:

(a) High privacy behavior but low concern for privacy and literacy. This situation is paradoxical (since high literacy should be consistent with behavior and concern), but it is rare and does not represent the classic privacy paradox.
(b) Low privacy behavior and privacy concern, with high literacy. There is no paradox regardless of the literacy level.
(c) High privacy behavior but low concern for privacy and low literacy. This situation is not paradoxical or presents an imaginary paradox (since low literacy explains the deviation).

(d) Low privacy behavior and concern for privacy with low literacy. There is no paradox regardless of the literacy level.
(e) Low privacy behavior but high concern for privacy and low literacy. This situation is not paradoxical or presents an imaginary paradox (since low literacy explains the deviation).
(f) High privacy behavior and high concern for privacy with low literacy. There is no paradox regardless of the literacy level.
(g) Low privacy behavior and high concern for privacy and high literacy. This is a classic example of the privacy paradox where privacy behavior deviates downwards from privacy concern despite high literacy level, so the deviation cannot be explained.
(h) High privacy behavior and high concern for privacy with high literacy. There is no paradox regardless of the literacy level.

13.3 Applying Soft Logic to the Privacy Paradox

To represent the privacy paradox with the mathematics of Soft logic, let C be *privacy concern*, let B be *privacy behavior*, and let L be *technological literacy*, such that $L \in [0, 1]$, where $L = 0$ indicates no literacy at all and $L = 1$ indicates maximal literacy. The expression $1 - L$ indicates the level of lack of literacy.

Let $\Delta = |C - B|$ be a factor that measures the gap between privacy behavior and privacy concern. The value $\Delta \cong 0$ represents adequate behavior, i.e., when privacy concern is consistent with privacy behavior, and there is no significant gap between them. This value is expected when $0 \ll L$ (significant literacy); otherwise, we face a paradox. The value $0 \ll \Delta$ is expected (or may be explained) only when $L \to 0$ (minimal literacy). Absoluting the difference $C - B$ causes the loss of some information, whether the individual has a frivolous pattern ($B < C$) or a paranoid pattern ($C < B$).

We now introduce two representations for placing an individual in the privacy paradox space.

Representation A:

$$\psi_1 = \{|C - B| \cdot L\} \cdot \bar{0} + C \cdot \bar{1} = \{\Delta \cdot L\} \cdot \bar{0} + C \cdot \bar{1}$$

In this representation, ψ_1 stands for an individual state where C is the baseline and expresses the rational component of the number and the unexplained deviation from privacy concern ($|C - B| \cdot L$) is the "Soft" component of the number. These two components may also be thought of as the expectancy and standard deviation.

In a world where the privacy paradox does not exist, the "Soft" component of ψ_1 is redundant, and ψ_1 is actually rational. In this world, $\Delta = f(L)$, where $f' < 0$.

However, in the real world, because $\Delta \neg \ \Box f(L)$ (Δ is not necessarily a function of L), the paradox does exist.

Representation B:

$$\psi_2 = \{|C - B| \cdot L\} \cdot \bar{0} + \{|C - B| \cdot (1 - L)\} \cdot \bar{1} = \{\Delta L\} \cdot \bar{0} + \{\Delta \cdot (1 - L)\} \cdot \bar{1}$$

In this representation, ψ_2 stands for an individual deviation level, where the unexplained deviation from privacy concern ($\{|C - B| \cdot L\}$) is the "Soft" component of the number and the explained deviation from privacy concern ($|C - B| \cdot (1 - L)$) is the rational component of the number.

To conduct the empirical research, we developed a questionnaire aimed at collecting information from individuals on the three main factors: privacy concerns (C), privacy behavior (B), and technological literacy (L). The empirical research combined privacy concerns and privacy behavior in seven scenarios:

(a) Using secure sites when purchasing products in an online transaction. A screenshot depicting this scenario was presented to participants.
(b) Posting on the Facebook social network about a car accident experienced by the individual.
(c) Consenting to provide personal information in response to a national survey. A screenshot depicting this scenario was presented to participants.
(d) An offer to receive a significant discount on electronics purchases in return for providing an email address. A screenshot depicting this scenario was presented to participants.
(e) Browsing a website by clicking on a URL link provided as a message stating that free movie tickets were won.
(f) The importance of updating antivirus software and insisting on an original license.
(g) Reading the privacy statement when installing a new application on a smartphone.

For each of the scenarios above, a pair of questions was presented, one to assess the actual action of respondents in that scenario (B) and the second to assess their attitude toward the scenario (C). The two parameters B and C were rated on a scale of 1–5. The answers were normalized for all scenarios, so that 1 represents the minimum value and 5 represents the maximum. To achieve this, the scale was flipped for some of the questions (according to context); for example, in scenario (b) on posting about a car accident, choosing the "public" publishing option represents a minimum value of 1, while the response of not publishing at all represents a maximum value of 5.

13.4 Results and Conclusions

Figure 13.2 describes the results of the questionnaires on the three tested parameters: attitude to privacy, privacy behavior, and technological literacy.

Figure 13.3 presents the Soft numbers defined using the models we developed. We present the Soft numbers in the Cartesian axis system and the Soft axis system.

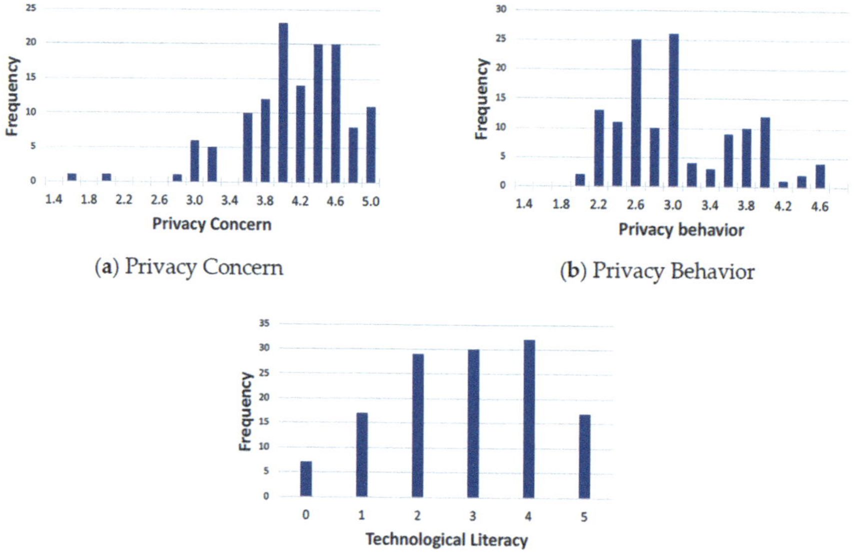

(a) Privacy Concern (b) Privacy Behavior

Fig. 13.2 Distributions of privacy concerns, privacy behavior, and technological literacy

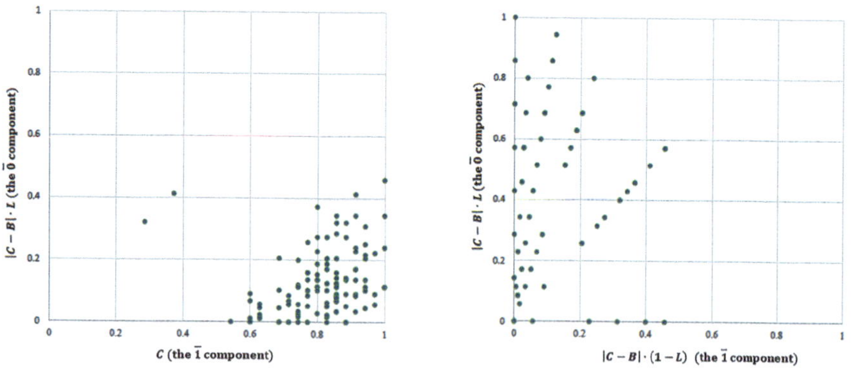

(a) Representation A (ψ_1) (b) Representation B (ψ_2)

Fig. 13.3 Graphical representations of the Soft numbers yielded by the empirical study

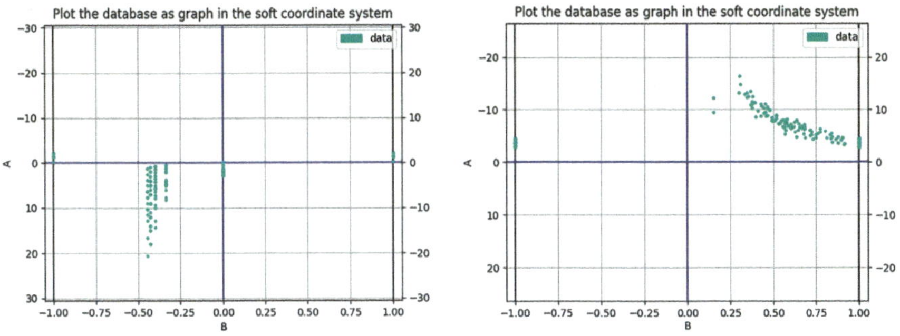

Fig. 13.4 Presentation of the data in the Soft coordinate system

As seen in Figure 13.4, presenting the same data in the Soft axis system creates a clear pattern or order that cannot be seen in the Cartesian system. This initial result testifies to the possibility and potential to describe mental states using Soft logic and Soft numbers. In this way, we may be able to understand a person's mental state more profoundly and propose effective tools to assist him.

This is only a preliminary study, and it is necessary to continue to investigate and establish the idea that Soft logic can be used to better understand peoples' mental behavior and improve the ability to help them.

References

Acquisti, A., Brandimarte, L., Loewenstein, G.: Privacy and human behavior in the age of information. Science. **347**, 509–514 (2015)

Anic, I.D., Budak, J., Rajh, E., Recher, V., Skare, V., Skrinjaric, B.: Extended model of online privacy concern: what drives consumers' decisions? Online Inf. Rev. **43**(5), 799–817 (2019). https://doi.org/10.1108/OIR-10-2017-0281

Arpetti, J., Delmastro, M.: The privacy paradox: a challenge to decision theory? J. Ind. Bus. Econ. **48**, 505–525 (2021). https://doi.org/10.1007/s40812-021-00192-z

Baek, Y.M.: Solving the privacy paradox: a counter-argument experimental approach. Comput. Hum. Behav. **38**, 33–42 (2014)

Cover, T.M., Thomas, J.A.: Elements of Information Theory, 2nd edn. Wiley-Interscience, Hoboken, NJ (2006). https://doi.org/10.1002/047174882X

Hirschprung, R.S., Klein, M., Maimon, O.: Harnessing Soft logic to represent the privacy paradox. Informatics. **9**(3), 54 (2022). https://doi.org/10.3390/informatics9030054

Chapter 14
Application Examples and Further Research

In this chapter, we summarize and conclude the foundation of Soft logic. We have presented a rigorous theory that is based on five axioms, which now enables us to develop several domains for investigation. We believe that this is just the beginning of a large area of further investigation and a solid basis for new scientific achievements. The chapter opens with an application of Soft logic to the representation of systemic thinking. For this purpose, we developed a linear algebra with Soft numbers. Later we describe how systemic thinking properties can be represented using matrices with Soft numbers. Later in the chapter, we develop the field of Soft analysis using a different approach than in Chap. 9. We define a distance between two Soft numbers and as a result present definitions of limits of sequences of Soft numbers and limits of Soft functions. We define derivatives of Soft function, and then we show how Pascal's triangle in combinatorics can be expanded using Soft numbers. We conclude the chapter by summarizing and pointing to additional possible developments using Soft logic.

14.1 Adopting Soft Logic to Systems Thinking

This section describes the extension of linear algebra with Soft logic, after which we describe a possible application to the domain of systems thinking (Hirschprung et al. 2023). First, we describe the application to linear algebra. As defined above, $\mathbf{SN} = \{a\bar{0}\dot{+}b, \quad a, b \in R\}$. Now, let a Soft vector be defined as follows: $v = (v_1, v_2, \ldots, v_n)$, when $v_i = u_i\bar{0}\dot{+}w_i \in \mathbf{SN}$. The presentation of a Soft vector can be split: $v = (u_1, \ldots u_n)\bar{0}\dot{+}(w_1, w_2, \ldots w_n)$ where $u_i \in R$, $w_j \in R$.

The addition operator on the two Soft vectors $v = (v_1, v_2, \ldots, v_n)$ and $u = (u_1, u_2, \ldots, u_n)$ will be defined as $v + w = (v_1 + u_1, v_1 + u_2 \ldots, v_n + u_n)$.

The multiplication operator of a Soft vector v (as defined above) with a scalar k ($k \in R$ or $k \in \mathbf{SN}$) (real or Soft) will be $kv = (kv_1, kv_2, \ldots, kv_n)$.

M. Klein, O. Maimon, *Foundations of Soft Logic*,
https://doi.org/10.1007/978-3-031-58233-2_14

Now, a Soft matrix can be defined as $C = \{c_{i,\,j} | c_{i,\,j} \in \mathbf{SN}\}$ when $1 \leq i \leq n$, $1 \leq j \leq m$, and each element of the matrix C can be split:

$$c_{i,j} = a_{i,j}\bar{0} \dot{+} b_{i,j}$$

$$\mathrm{A} = \{a_{i,j}\} \quad a_{i,j} \in R \quad 1 \leq i \leq n; \quad 1 \leq j \leq m$$

$$\mathrm{B} = \{b_{i,j}\} \quad b_{i,j} \in R \quad 1 \leq i \leq n; \quad 1 \leq j \leq m$$

Therefore, C can have a canonical presentation $C = A\bar{0} \dot{+} B$ (where A and B are real matrices):

$$C = \begin{Bmatrix} c_{1,1} & c_{1,2} & \cdots & c_{1,n} \\ c_{2,1} & c_{2,2} & \cdots & c_{2,n} \\ \vdots & \vdots & \ddots & \vdots \\ c_{n,1} & c_{n,1} & \cdots & c_{n,n} \end{Bmatrix}$$

The addition operator of two Soft matrices will be

$$C_1 = A_1\bar{0} \dot{+} B_1 \text{ and } C_2 = A_2\bar{0} \dot{+} B_2 \text{ is}$$

$$C_1 + C_2 = (A_1 + A_2)\bar{0} \dot{+} (B_1 + B_2)$$

The multiplication operator of two Soft matrices will be

$$C_1C_2 = (A_1B_2 + B_1A_2)\bar{0} \dot{+} B_1B_2$$

Since the multiplication of matrix is not commutative, the power of the Soft matrix

$$C = A\bar{0} \dot{+} B :$$

is

$$\mathrm{C}^2 = (\mathrm{AB} + \mathrm{BA})\bar{0} \dot{+} B^2$$

And in general

$$C^n = \left(\sum_{k=0}^{n-1} B^k A B^{n-k}\right)\bar{0} \dot{+} B^n$$

which can be proved by induction. For $n = 1$ is trivial, for $n = 2$

$$C^2 = CC = \left(\underbrace{AB + BA}_{= B^0AB^1 + B^1AB^0}\right)\bar{0} \dot{+} B^2 = \left(\sum_{k=0}^{2-1} B^k A B^{2-k}\right)\bar{0} \dot{+} B^2$$

If true for n, then true for $n + 1$:

$$C^{n+1} = C^nC = \left[\left(\sum_{k=0}^{n-1} B^kAB^{n-k}\right)\bar{0} \dot{+} B^n\right](A\bar{0} + B)$$

$$= \left(\sum_{k=0}^{n-1} B^kAB^{n-k}B^1 + B^nAB^0\right)\bar{0} + B^{n+1}$$

$$= \left(\sum_{k=0}^{n} B^kAB^{n+1-k}\right)\bar{0} \dot{+} B^{n+1}$$

The inverse of a Soft matrix is

$$(A\bar{0} \dot{+} B)^{-1} = -B^{-1}AB^{-1}\bar{0} \dot{+} B^{-1}$$

Because

$$\text{Let } (A\bar{0} + B)^{-1} = X\bar{0} + Y, \text{ then}$$

$$(A\bar{0} \dot{+} B)(X\bar{0} \dot{+} Y) = (AY + BX)\bar{0} \dot{+} BY = O\bar{0} + I$$

$$(X\bar{0} \dot{+} Y)(A\bar{0} \dot{+} B) = (YA + XB)\bar{0} \dot{+} YB = O\bar{0} + I$$

Where I is the identity matrix and O the zero matrix.
From the real part:

$$YB = BY = I \Rightarrow Y = B^{-1}$$

From the zero part:

$$AY + BX = YA + XB = O \Rightarrow X = -B^{-1}AB^{-1}$$

The Soft matrices C can be presented as

$$C = \begin{Bmatrix} \overrightarrow{c_1} \\ \vdots \\ \overrightarrow{c_n} \end{Bmatrix}$$

Now, let $\vec{w}$ be a weights vector that provides different weights to the different dimensions $\vec{w} = (w_1, w_2, \ldots, w_n), \; w_k \in R$.

Weighting C by $\vec{w}$, which is actually the multiplication of Soft matrices with a real weight vector, is represented by

$$C\vec{w} = \begin{Bmatrix} \vec{c}_1\vec{w} \\ \vdots \\ \vec{c}_n\vec{w} \end{Bmatrix},$$

where

$$\vec{c}_1\vec{w} = (a_{1,1}\bar{0} + b_{1,1})w_1 + (a_{1,2}\bar{0} + b_{1,2})w_2 + \ldots (a_{1,n}\bar{0} + b_{1,n})w_n$$
$$\vec{c}_k\vec{w} = (a_{k,1}\bar{0} + b_{k,1})w_1 + (a_{k,2}\bar{0} + b_{k,2})w_2 + \ldots (a_{k,n}\bar{0} + b_{k,n})w_n$$
$$\vec{c}_n\vec{w} = (a_{n,1}\bar{0} + b_{n,1})w_1 + (a_{n,2}\bar{0} + b_{n,2})w_2 + \ldots (a_{n,n}\bar{0} + b_{n,n})w_n$$

The order of the two Soft numbers $c_1 = a_1\bar{0}\dot{+}b_1$ and $c_2 = a_2\bar{0}\dot{+}b_2$ is defined by a lexicographic approach. First, we compare the real part, and then, if necessary, we compare the Soft part but in the opposite direction. The formal definition is

$$c_1 < c_2 \Leftrightarrow (b_1 < b_2) \text{ or } ((b_1 = b_2) \text{ and } (a_2 < a_1))$$

Now we describe a possible application in the domain of systems thinking. Systems thinking involves a process of exploration and understanding and of delving into the underlying mechanisms that shape the challenges we encounter and the opportunities we have. It provides a framework for comprehending the forces and interconnectedness that influence the behavior of systems.

We expressed and measured individual systems thinking skills through four dimensions: {knowledge, individual traits, cognitive characteristics, capabilities}, marked as $S = \{s_1, s_2, s_3, s_4\}$, respectively. For a specific individual, each dimension s_i may receive a continuous value on a predetermined scale, which represents the level of this dimension (the higher the better) regarding this individual. While the literature on systems thinking suggests these four dimensions, the model is flexible enough to accommodate any discrete number of dimensions of other proficiencies or skills.

The core issue is the fact that different people do not line up and correlate with regard to the elements of S. For example, imagine two people, one with $S_1 = \{9, 5, 3, 7\}$ and the other with $S_2 = \{9, 3, 5, 7\}$; it can be noticed that the first one is stronger in s_2, while the second one is stronger in s_3.

14.1.1 Representation of Systems Thinking

Considering the combination of each two dimensions, the central values of these dimensions, which are their averages, are expressed in the matrix $S^{\bar{1}}$. Each row and column index of the matrix $S^{\bar{1}}$ are formed by S. The diagonal represents the combination of each dimension with itself and therefore the sole dimension. The values above the diagonal represent the average of two dimensions, while the values below the diagonal are discarded, as they are a mirror image of the ones above. In the Soft logic representation, this matrix stands for the $\bar{1}$ component of the Soft number. Semantically, the matrix $S^{\bar{1}}$ represents the *central values* of the dimensions, ignoring the differences between them.

$$S^{\bar{1}} = \left(\frac{s_i + s_j}{2}\right) = \begin{pmatrix} s_1 & \frac{s_1 + s_2}{2} & \cdots & \cdots \\ \cdot & \ddots & \vdots & \vdots \\ \cdot & \cdot & \ddots & \ddots \\ \cdot & \cdot & \cdot & \ddots \end{pmatrix} \tag{14.1}$$

The differences between the dimensions span a combinatorial space of the combinations of each two dimensions (naturally, order is neglected) and are expressed in the matrix $S^{'}$. Each row and column are indexed in the same way as in (14.1), and each value above the diagonal is the distance (therefore the expression is based on the absolute value) between the average value expressed in (14.1) and each of the values (which is the same for both of them). The diagonal represents the combination of each dimension with itself and therefore must be 0, and as in (14.1), the values are mirrored, and therefore both are not mentioned.

$$S' = \left(\left|s_i - s_j\right|\right) = \begin{pmatrix} 0 & |s_1 - s_2| & |s_1 - s_3| & |s_1 - s_4| \\ \cdot & 0 & |s_2 - s_3| & |s_2 - s_4| \\ \cdot & \cdot & 0 & |s_3 - s_4| \\ \cdot & \cdot & \cdot & 0 \end{pmatrix} \tag{14.2}$$

However, to handle small and insignificant differences, we introduce a threshold, in which below this value the difference is set to *zero*; thus, the complete differences matrix $S^{\bar{0}}$ is defined as shown in (14.3). In the Soft logic representation, this matrix stands for the $\bar{0}$ component of the Soft number. Semantically, the matrix $S^{\bar{0}}$ represents the *tension* between the dimensions.

$$S^{\bar{0}} = \left(\begin{cases} \frac{|s_i - s_j|}{2} & |s_i - s_j| > T \\ 0 & \text{otherwise} \end{cases}\right) = \begin{pmatrix} \cdot & \begin{cases} \frac{|s_1 - s_2|}{2} & |s_1 - s_2| > T \\ 0 & \text{otherwise} \end{cases} & \cdots & \cdots \\ \cdot & \cdot & \vdots & \vdots \\ \cdot & \cdot & \cdot & \ddots \\ \cdot & \cdot & \cdot & \cdot \end{pmatrix} \tag{14.3}$$

Finally, the comprehensive ranking of an individual's systems thinking skills is expressed by *STS*, as shown in (14.4). *STS* is a Soft number that encapsulates both the central values of the dimensions and the tensions between them.

$$\text{STS} = S^{\bar{0}} \otimes \bar{0} \dotplus S^{\bar{1}} \otimes \bar{1} \tag{14.4}$$

14.1.2 Ordering

Assume we have two individuals with systems thinking levels of STS_1 and STS_2, respectively. In addition, the recruiter weighted the dimensions of S with the vector $\vec{w} = (w_1, w_2, \ldots, w_n)$, where $w_k \in R$ and $\sum_{\forall i} w_i = 1$. Each element of w_i refers to the parallel element of S and indicates its weight. The overall rank of the central values of STS, denoted by $\text{STS}^{r\bar{1}}$, is the sum of the multiplications of each member of the diagonal of matrix $S^{\bar{1}}$ with the corresponding member of $\vec{w}$:

$$\text{STS}^{r\bar{1}} = \sum_{\forall i} S^{\bar{0}}{}_{i,i} \cdot w_i$$

Also, the overall rank of the tensions of STS, denoted by $\text{STS}^{r\bar{0}}$, is the sum of the multiplication of each member of the matrix $S^{\bar{0}}$ by the average of the corresponding members of $\vec{w}$:

$$\text{STS}^{r\bar{0}} = \sum_{\forall i \text{ and } j} S^{\bar{0}}{}_{i,j} \cdot \frac{w_i + w_j}{2}$$

Now, STS_1 and STS_2 can be ordered according to the rule derived from the equation:

$$\text{STS}_1 < \text{STS}_2 \Leftrightarrow \left(\text{STS}^{r\bar{1}}{}_1 < \text{STS}^{r\bar{1}}{}_2\right) \text{ or } \left(\left(\text{STS}^{r\bar{1}}{}_1 = \text{STS}^{r\bar{1}}{}_2\right) \text{ and } \left(\text{STS}^{r\bar{0}}{}_2 < \text{STS}^{r\bar{0}}{}_1\right)\right)$$

This is the model that we developed with Soft logic and suggest applying to the presentation of systems thinking.

14.2 Soft Logic Analysis

In this section, we present a new definition of the distance between two Soft numbers, based on the definition of the height of a Soft number in the Soft coordinate system and the Manhattan distance. The metric space axioms are obtained by this new Soft number distance.

With the new definition of a distance between two Soft numbers, we have the ability to define the limit of a sequence of Soft numbers and Soft functions. We define a Soft analytic function and derive the result of a limit of a Soft analytic function as a special case. Based on the limit of a Soft function, we define the derivative of a Soft function.

14.2.1 Our Motivation

Based on the preliminary definitions and axioms, we wish to find a way to measure the distance between two Soft numbers. Based on such a definition of distance, we will be able to write a definition for Soft number sequences that converge to a Soft number, i.e., limits of Soft number sequences. In addition, we will be able to define limits of functions that are applied to Soft numbers, derivatives, and integrals. In Appendix A, we demonstrate the properties of applying elementary functions (exponential, logarithmic, and trigonometric functions) to Soft numbers.

Definition 14.1 The distance between two Soft numbers. Suppose we have two Soft numbers $z_1 = x_1\bar{0}\dot{+}y_1$ and $z_2 = x_2\bar{0}\dot{+}y_2$, then the distance between them is defined as follows: I. $d(z_1, z_2) = |z_1 - z_2| = |x_1 - x_2| + |y_1 - y_2|$. Due to the structure of the taxicab/Manhattan distance on the right-hand side of the definition, this definition satisfies the axioms of a metric space. Recall the definition of a metric space:

Definition 14.2 (Metric space). Suppose Z is a set and $d : Z \times Z \to [0, \infty)$ is a function that satisfies the following properties for all z_1, z_2, $z_3 \in Z$.

(1) Positive property: $d(z_1, z_2) \geq 0$ with equality iff (if and only if) $z_1 = z_2$.
(2) Symmetric property: $d(z_1, z_2) = d(z_2, z_1)$.
(3) Triangle inequality: $d(z_1, z_2) \leq d(z_1, z_3) + d(z_2, z_3)$.

Then d is called a "metric" on Z and the ordered pair (Z, d) is called "a metric space."

Lemma 14.1 The ordered pair (SN, d), where d is the distance function is a metric space.

14.2.2 Limits of Soft Number Sequences

In this section, we define the criterion for a sequence of Soft numbers to converge to a Soft number $\{c_n = a_n\bar{0} + b_n\}_{n=1}^{\infty}$, where $c = a\bar{0}\dot{+} b$. We use the same definition of a limit of a sequence of real numbers to converge to a real number.

Definition 14.3 We say that the sequence of Soft numbers converges to a Soft number c, i.e.,

$$\lim_{n\to\infty} c_n = c$$

$$\forall \varepsilon > 0, \exists N \in N, \forall n \in N, n > N \Rightarrow \mid c_n - c \mid < \varepsilon.$$

In the above context of the definition, ε is a positive real number, as small as we want.

Theorem 14.1 Let $c_n = a_n\bar{0} + b_n$ be a sequence of Soft numbers and $c = a\bar{0}\dot{+} b$. Then the sequence of the Soft numbers c_n converges to a Soft number c iff each of its components of c_n (an in the zero axis and b_n in the 1 axis) converge to the counterpart of c's components (a in the zero axis and b in the 1 axis), i.e., $\lim_{n\to\infty} c_n = c \Leftrightarrow \lim_{n\to\infty} a_n = a,\ \lim_{n\to\infty} b_n = b.$

14.2.3 Limits of Soft Functions

Let $\bar{f} : SN \to SN$ be a general function with the form $\bar{f}(x\bar{0}\dot{+}y) = f_0(x, y)\bar{0}\dot{+}f_1(x, y)$.

Where $f_{0,\,1} : \mathbb{R}^2 \to \mathbb{R}$, we say that $\bar{f}(x\bar{0}\dot{+}y)$ is a Soft analytic function if $\bar{f}(x\bar{0}\dot{+}y)$ satisfies the following equations:

$$f_1(x, y) = f(y)$$
$$f_0(x, y) = xf'(y)$$
$$\Rightarrow \bar{f}(x\bar{0}\dot{+}y) = xf'(y)\bar{0}\dot{+}f(y)$$

This definition extends a real differentiable function to a Soft function $\bar{f}$. We say that the Soft function $f(z),\ z = x\bar{0}\dot{+}y$ converges to a Soft number $L = L_0\bar{0}\dot{+}L_1$ as z approaches to $c = z = a\bar{0}\dot{+}b$.

$$\lim_{z\to c} f(z) = L,$$

iff

$$\forall \varepsilon > 0, \exists \delta > 0, \forall z \in SN, \quad |z - c| < \delta \Rightarrow \mid f(z) - L \mid < \varepsilon$$

Or equivalently

$$|x - a| + \mid y - b \mid < \delta \Rightarrow \mid f(x\bar{0}\dot{+}y) - (L_0\bar{0}\dot{+}L_1) \mid < \varepsilon$$

14.2.4 Derivatives of Soft Functions

In this section, we develop the derivatives of Soft functions based on the notion of derivatives as defined in real analysis but using the limits of Soft functions. Suppose that we want to calculate the rate change of the Soft function$f(z)=f_0(x,y)\bar{0}\dot{+}f_1(x,y),\ \ z=x\bar{0}\dot{+}y$ at the point $c=a\bar{0}\dot{+}b$.

Let the following differences be defined as follows:

$$x-a=\Delta x$$

$$y-b=\Delta y$$

$$z-c=\Delta z=\Delta x\bar{0}\dot{+}\Delta y$$

$$f(z)-f(c)=\Delta f=\Delta f_0\bar{0}\dot{+}\Delta f_1$$

After applying these equations, we have the following expression for the reciprocal of Δz:

$$\frac{1}{\Delta z}=-\frac{\Delta x}{\Delta y}\cdot\frac{1}{\Delta y}\bar{0}\dot{+}\frac{1}{\Delta y}$$

$$\frac{\Delta f}{\Delta z}=\left(\frac{\Delta f_0}{\Delta y}-\frac{\Delta x}{\Delta y}\cdot\frac{\Delta f_1}{\Delta y}\right)\bar{0}\dot{+}\frac{\Delta f_1}{\Delta y}$$

$$f_i(x,y)\approx f_i(a,y)+\frac{\partial f_i(a,y)}{\partial x}(x-a)$$

$$\frac{\Delta f_i}{\Delta y}=\frac{\partial f_i(a,b)}{\partial y}+\frac{\partial f_i(a,b)}{\partial x}X'(b)$$

where $X'(b)$ is a derivative of the curve $x=X(y)$ that satisfies $X(b)=a$.

Finally, we can write the expression of the derivative of f

$$f'\left(x\bar{0}\dot{+}y\right)=\left(\frac{df_0}{dy}-X'(y)\right)\bar{0}\dot{+}\frac{df_1}{dy}$$

$$\frac{df_i}{dy}=\frac{\partial f_i}{\partial y}+\frac{\partial f_i}{\partial x}X'(y),\quad x=X(y)$$

In the case that x and y are independent, the derivative $X'(y)$ vanishes, so that the derivative $f'\left(x\bar{0}\dot{+}y\right)$ is simplified to $f'\left(x\bar{0}\dot{+}y\right)=\frac{\partial f_0}{\partial y}\bar{0}\dot{+}\frac{\partial f_1}{\partial y}$.

In case f is analytic, we have the following derivative: $f'\left(x\bar{0}\dot{+}y\right)=xf''(y)\bar{0}\dot{+}f'(y)$.

14.3 Extending the Pascal Triangle with Soft Numbers

In this section, we present an extension of the known Pascal triangle with Soft numbers (Klein and Maimon 2023). Given a Soft number, $C = X\bar{0} \dot{+} Y$, where $XY > 0$,we can measure its level of distinction by the distance to the zero line: $B = \frac{Y}{X+Y}$.

By this definition, we can easily see that a number on the one line ($X = 0$) has distinction y 1 and a number on the zero axis ($Y = 0$) has distinction 0.

Now the entropy of the Soft number C can be calculated by

$$H(C) = -B\mathrm{Log}(B) = -\frac{Y}{X+Y}\log\left(\frac{Y}{X+Y}\right)$$

We wish to investigate the case of $Y = 0$. To do so, we need to develop a probability theory of the zero axis. The question is how to deal mathematically with events whose probability is zero? Let us look at the binomial distribution

$$\Pr(k, n, p) = \binom{n}{k} p^k (1-p)^k.$$

For $k = 0, 1, 2, \ldots n$, the binomial coefficients are $C_k^n = \frac{n!}{k!(n-k)!}$.

These numbers can be arranged in the known Pascal triangle, Fig. 14.1.

The general property of the triangle is $\binom{n}{k} + \binom{n}{k+1} = \binom{n+1}{k+1}$.

Can we extend the computation of the coefficient when $k > n$? See Fig. 14.2.

Following the observation of Abraham Senkar, we extend the Pascal triangle with Soft numbers by the following procedure: It is known that $0! = 1$. The recursion property of the factorial function is $(n-1)! = \frac{n!}{n}$, so we can assume that

$$(-1)! = \frac{0!}{0} = \frac{1}{0} \equiv \frac{1}{\bar{0}}$$

and

Fig. 14.1 The Pascal triangle

1
1 1
1 2 1
1 3 3 1
1 4 6 4 1
1 5 10 10 5 1

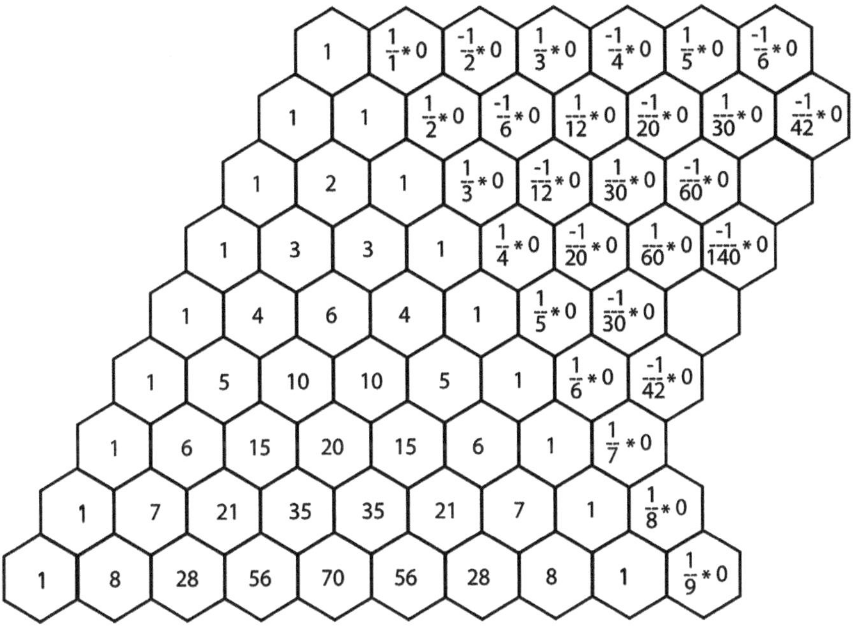

Fig. 14.2 The extended Pascal triangle

$$(-2)! = \frac{(-1)!}{-1} = -\frac{1}{\bar{0}}$$

and

$$(-3)! = \frac{(-2)!}{-2} = \frac{1}{2\bar{0}}$$

and in general

$$(-m)! = \frac{(-(m-1))!}{-(m-1)} = (-1)^{m-1}\frac{1}{(m-1)!\bar{0}}$$

Therefore, the general extension of a Pascal triangle with Soft numbers is

$$\binom{n}{k} = \frac{n!}{k!(n-k)!} = \frac{n!}{k!(-(k-n))!} = (-1)^{k-n-1}\frac{n!(k-n-1)!}{k!}\bar{0}$$

Theorem 14.2 For k>n, the extension of a Pascal triangle satisfies the usual condition $\binom{n}{k} + \binom{n}{k+1} = \binom{n+1}{k+1}$.

Proof 14.2

$$(-1)^{k-n-1}\frac{n!(k-n-1)!}{k!}\overline{0}+(-1)^{k-n}\frac{n!(k-n)!}{(k+1)!}\overline{0}=$$

$$=(-1)^{k-n-1}\frac{n!(k-n-1)!}{k!}\left(1-\frac{(k-n)}{(k+1)}\right)\overline{0}=$$

$$=(-1)^{k-n-1}\frac{n!(k-n-1)!}{k!}\frac{(n+1)}{(k+1)})\overline{0}$$

$$=(-1)^{k-n-1}\frac{(n+1)!(k-n-1)!}{(k+1)!}\overline{0}=\binom{n+1}{k+1}$$

■

14.4 Conclusions and Further Research

We have shown that the concept of the infinitesimal can be expressed in a tangible way and not as an expression of a limit. In addition, we have shown that Soft numbers are found on the surface of a Möbius strip and thus express a bridge between internal and external reality. We believe that the very existence of Soft numbers is an expression of the organic unity of mathematics.

We have shown the connection between Soft numbers and the geometry of the Möbius strip and the connection to the general theory of calculus. We have also shown the dynamics of Soft functions and the connection to the theory of fractals. We showed that it is possible to develop Soft analysis and extend the known Möbius function in number theory to a Soft Möbius function. We applied Soft numbers to the theory of probability. Also, we extended the known Pascal triangle with Soft numbers. With these cases, we suggested a way to observe the unity of mathematics.

We believe that the unity of mathematics derives from the ability to observe creation as whole, mathematics as one living, organic essence. This approach is in contrast to the usual specialization process of choosing a particular field or industry and focusing on trying to solve a specific open problem. The intention in the question of the unity of mathematics is to try to create a new point of view on all mathematical fields in such a way that is not only philosophical but that brings forth a specific mathematical practice.

In the twentieth century, several unsuccessful attempts were made to unify mathematics through ordinary logic. Soft logic offers a new approach to dealing with the question of the unity of mathematics. According to our understanding, the unity of mathematics is achieved as soon as the creative process of mathematics becomes an organic part of the products of mathematics. In this approach, there is really no separation between mathematics and the mathematician who discovers and creates it. In order to reach this understanding, we will be required to unite the mathematical results with the rate of their change and creation (such as in Chap. 6).

The philosopher Pythagoras discovered the connection between music and mathematics. He taught his students that everything is numbers. In the seventh century, the number zero was invented, which greatly contributed to the development of mathematics and human culture. About 500 years ago, complex numbers were invented, and today they have many critical uses. Soft numbers are another extension of the number concept, and we can try to apply them to other areas of mathematics, such as combinatorics, differential equations, numerical analysis, and various other diverse branches of mathematics, physics and also in biology vis à vis the question of the origin of life (Torday et al. 2023).

References

Hirschprung, R.S., Kordova, S., Klein, M., Maimon, O.: Representation and assessment of systems thinking competencies through Soft logic. IEEE Access. **11**, 141547–141558 (2023). https://doi.org/10.1109/ACCESS.2023.3342131

Klein, M., Maimon, O.: Soft logic as an extension of Pascal's work. P-Adic. Num. Ultrametr. Anal. Appl. **15**, 119–132 (2023). https://doi.org/10.1134/S207004662302005X

Torday, J., Klein, M., Maimon, O.: The Möbius strip, the cell, and Soft Logic mathematics. Prog. Biophys. Mol. Biol. (2023)

Appendices

A.1 Nonstandard Analysis

Abraham Robinson developed nonstandard analysis in 1960. Shortly, his work extends the set of real numbers $\mathbb{R}$ to the set of hyperreals $\mathbb{R}^*$, which includes infinitesimal numbers. While Robinson relied on the compactness theorem of logic, others (including Luxemburg and Keislar) preferred a more direct construction of "ultrapowers." By them $\mathbb{R}^* = \frac{\mathbb{R}^\omega}{M}$, where M is an ideal of all infinite sequences that are equivalent to zero, under a chosen ultrafilter F. A number $\delta \epsilon \mathbb{R}^*$ is called infinitesimal if $-\varepsilon < \delta < \varepsilon$, for every real positive number ε.

In the ultrapower $\mathbb{R}^\omega$, infinite sequences describe the elements of $\mathbb{R}^*$. The standard real number c is identified with the sequence $(c, c, c, \ldots) \equiv c$. Two infinite sequences, $a = (a_1, a_2, a_3 \ldots)$, $b = (b_1, b_2, b_3 \ldots)$, are equal if the set of indices that have equal elements, i.e., $a_i = b_i$ is a big set. A big set is a member of an ultrafilter F, which in turn is a non-empty set of subsets of the natural numbers N that satisfies the following conditions:

1. *For all A, B, if* $A \in F$ *and* $A \subset B$, *then* $B \in F$.
2. *For all* $A_1, A_2 \in F \Rightarrow A_1 \cap A_2 \in F$.
3. *For all* $A \in P(N)$, *either* $A \in F$ *or* $A^c \in F$.

In Robinson's model of $\mathbb{R}^*$, the sequence $\delta = (1, \frac{1}{2}, \frac{1}{3}, \frac{1}{4}, \ldots \frac{1}{n}, \ \ldots)$ is an infinitesimal number because for any natural m there is $0 < \delta < \frac{1}{m}$. The disadvantage of Robinson's construction of $\mathbb{R}^*$ is that it uses non-constrictive tools. In Soft logic, we built a model for Soft zeros, which are in some way close to infinitesimals, using the zero line, without Zorn's lemma or the compactness theorem. In the following, we show how to represent Soft numbers using nonstandard analysis.

M. Klein, O. Maimon, *Foundations of Soft Logic*,
https://doi.org/10.1007/978-3-031-58233-2

How can we Represent bridge Numbers by Using a Model from nonstandard Analysis

In $\mathbb{R}^*$, there exists an infinitesimal $y > 0$, but there is a difference between y and $\overline{0}$ since $(\overline{0})^2 = 0$ but $y^2 \neq 0$. It is possible to treat y^2 as very small and neglect it and by that to connect it more strongly to $\overline{0}$. In order to do this, we define when a number δ is "very small" in relation to y.

Definition A1 An infinitesimal δ is neglected in relation to an infinitesimal y if $\frac{\delta}{y}$ is infinitesimal. We denote by I_y the set of all infinitesimals, which are neglected in relation to infinitesimal y.

Lemma A1 If y is an infinitesimal, then $y^2 \in I_y$.

Proof A1 Since $\frac{y^2}{y} = y$ and y are an infinitesimal, the statement is true.

We define the following set: $K = \{ay + b + \delta : a, b \in \mathbb{R}, \delta \in I_y\}$. K is a subset of $\mathbb{R}^*$ closed under addition and multiplication. We can define the mapping D of K on SN: K->SN, where $D(ay + b + \delta) = a\overline{0}\dot{+}b$ for any $\delta \in I_y$.

It can be shown that for all x, z in K, we get $D(x + z) = D(x) + D(z)$ and $D(x{*}z) = D(x){*}D(z)$. Therefore the mapping D is a homomorphism. Also, it has the following properties:

1. $D(y) = \overline{0}$.
2. For all x in K, we have $D(x) = 0$ if x is in I_y. This proves that $D(y){*}\,D(y) = 0$.
3. For all real x, $D(x) = x$.

The analogy of K to BN is as follows: Using D, numbers of the form ay+b+δ are mapped into $a\overline{0}\dot{+}b$ (numbers in I_y are mapped to 0, and y is mapped to $\overline{0}$). As we have $y^2 \in I_y$, it means that y^2 becomes 0, after the mapping.

A.2 Dual Numbers

The theory of dual numbers was developed by Clifford. A dual number is a number of the form ***a* + *bε***, where a and b are real numbers and $\varepsilon^2 = 0$.

Addition of dual numbers is defined as

$$(a + b\varepsilon) + (c + d\varepsilon) = (a + c) + (b + d)\varepsilon$$

Multiplication of dual numbers is defined as

$$(a + b\varepsilon) \times (c + d\varepsilon) = (ac) + (ad + bc)\varepsilon$$

It can be proven that

$$(a + b\varepsilon)^n = a^n + nba^{n-1}\varepsilon$$

Also, it can be proven that there is an algebraic isomorphism between dual numbers and bridge numbers: $\boldsymbol{a + b\varepsilon \Leftrightarrow b\bar{0} \dot{+} a}$.

The Ring of Polynomials

The ring of polynomials R[X] is the set of all finite polynomials over the real numbers under addition and multiplication:

$$R[X] = \left\{a_0 + a_1X + a_2X^2 + \ldots a_nX^n : X, a_i \in \mathbb{R}\right\}$$

The algebra of dual numbers is equivalent to the algebra of quotient ring:

$$\frac{R[X]}{< X^2 >} = \{a + bX : a, b, X \in \mathbb{R}\}$$

The isomorphism between bridge numbers and the quotient ring of polynomials is given by $\boldsymbol{a + bX \Leftrightarrow b\bar{0} \dot{+} a}$.

The main difference between the sets of Soft numbers and dual numbers and the quotient ring of polynomials is in their geometric interpretation. Soft logic is based on the existence of the zero axis, which is an extension of the number zero. With the zero axis we create a new coordinate system that is a simple model for the Möbius strip.

A.3 Fractals

The mathematician Felix Hausdorff first introduced *fractals,* a class of complex geometric shapes, in 1918. Fractals are distinct from the simple figures of classical, or Euclidean, geometry—the square, the circle, the sphere, and so forth. They are capable of describing many irregularly shaped objects or spatially nonuniform phenomena in nature such as coastlines and mountain ranges. The term *fractal*, derived from the Latin word *fractus* ("fragmented" or "broken"), was coined by the Polish-born mathematician Benoit B. Mandelbrot, who discovered the fractal set that is named today the Mandelbrot set (Fig. 8.1).

Mandelbrot was the first to point out that fractals could be an ideal tool in applied mathematics for modeling a variety of phenomena, from physical objects to the behavior of the stock market. Since its introduction in 1975, the concept of the fractal has given rise to a new system of geometry that has had a significant impact on such diverse fields as physical chemistry, physiology, and fluid mechanics.

Many fractals possess the property of self-similarity, at least approximately, if not exactly. A self-similar object is one whose component parts resemble the whole. This reiteration of details or patterns occurs at progressively smaller scales and can, in the case of purely abstract entities, continue indefinitely, so that each part of each part, when magnified, will look basically like a fixed part of the whole object. In effect, a self-similar object remains invariant under changes of scale, i.e., it has scaling symmetry. This fractal phenomenon can often be detected in objects such as snowflakes and tree barks. All natural fractals of this kind, as well as some mathematical self-similar ones, are stochastic or random; they thus scale in a statistical sense.

Another key characteristic of a fractal is a mathematical parameter called its fractal dimension. Unlike Euclidean dimensions, fractal dimensions are generally expressed by a non-integer, that is to say, by a fraction rather than by a whole number.

Fractal geometry, with its concepts of self-similarity and non-integer dimensionality, has been applied increasingly in statistical mechanics, notably when dealing with physical systems consisting of seemingly random features. For example, fractal simulations have been used to plot the distribution of galaxy clusters throughout the universe and to study problems related to fluid turbulence. Fractal geometry also has contributed to computer graphics. Fractal algorithms have made it possible to generate lifelike images of complicated, highly irregular natural objects, such as the rugged terrains of mountains and the intricate branch systems of trees.

Mathematical Definition of the Mandelbrot Set

We define the following sequences for any complex number c:

$$a_0(c) = 0$$

$$a_{n+1}(c) = a_n(c)^2 + c$$

The Mandelbrot set is the set of all complex numbers c for which the sequence $a_n(c)$ does not diverge to infinity when iterated from $z = 0$. This definition can be extended to other functions $f(c)$ $a_{n+1}(c) = f(c)$, thereby generating other types of fractals.

In the dynamic of Soft numbers we define similar process, but instead of complex numbers we apply Soft functions:

$$f(c) = f\left(x\bar{0} \dot{+} y\right)$$

$$a_0 = c$$

$$a_{n+1} = f(a_c)$$

Like fractals, we distinguish between the limit and infinity but also between the zero axis and the one axis, thereby generating a new Soft type of fractals.

A.4 The Riemann Hypothesis

The Riemann hypothesis, in number theory, is the hypothesis formulated by the German mathematician Bernhard Riemann concerning the location of solutions to the Riemann zeta function, which is connected to the prime number theorem and has important implications for the distribution of prime numbers. The zeta function is defined as the infinite series: $\zeta(s) = \sum_{n=1}^{\infty} \frac{1}{n^s}$, where the summation ($\Sigma$) of terms for n runs from 1 to infinity through all the positive integers and s is a fixed positive integer greater than 1. The zeta function was first studied by the Swiss mathematician Leonhard Euler in the eighteenth century. In 1737, Euler discovered a formula relating the zeta function, which involves summing an infinite sequence of terms containing all positive integers and an infinite product involving every prime number:

$$\zeta(s) = \sum_{n=1}^{\infty} \frac{1}{n^s} = \prod_p \frac{1}{1 - p^{-s}}$$

Riemann extended the study of the zeta function to include the complex numbers $x + iy$, where $i =$ the square root of $\sqrt{-1}$, except for the line $x = 1$ in the complex plane. Riemann knew that the zeta function equals zero for all negative even integers $-2, -4, -6, \ldots$ (so-called trivial zeros) and that it has an infinite number of zeros in the critical strip of complex numbers that fall strictly between the lines $x = 0$ and $x = 1$. He also knew that all nontrivial zeros are symmetric with respect to the critical line $x = 1/2$. Riemann conjectured that all of the nontrivial zeros are on the critical line, a conjecture that subsequently became known as the Riemann hypothesis.

In 1914, the English mathematician Godfrey Harold Hardy proved that an infinite number of solutions of $\zeta(s) = 0$ exist on the critical line $x = 1/2$. Subsequently, it was shown by various mathematicians that a large proportion of the solutions must lie on the critical line, though the frequent "proofs" that all the nontrivial solutions are on it have been flawed. Computers have also been used to test solutions, with the first 10 trillion nontrivial solutions shown to lie on the critical line.

A proof of the Riemann hypothesis would have far-reaching consequences for number theory and for the use of primes in cryptography. The Riemann hypothesis has long been considered the greatest unsolved problem in mathematics. It was one of 10 unsolved mathematical problems (23 in the printed address) presented as a challenge for twentieth-century mathematicians by the German mathematician David Hilbert at the Second International Congress of Mathematics in Paris on August 8, 1900.

It has been proven that there are deep connections between the Riemann hypothesis and the Möbius function in number theory ($\mu(n) \in \{-1, 0, 1\}$). We extend and define the Soft Möbius function $\overline{\mu}(n) \in \{-0, 0, +0, -1, 1\}$ and prove a connection to the Riemann hypothesis.

The Mertens function is defined by $\mathrm{M}(x) = \sum_{n \leq x} \mu(n)$. It is known that the Riemann hypothesis is equivalent to that for every $\varepsilon > 0$, $\lim_{x \to \infty} \frac{M(x)}{x^{\frac{1}{2}+\varepsilon}} = 0$. We prove that this equivalence still holds with the extension to the Soft Möbius function.

GPSR Compliance

The European Union's (EU) General Product Safety Regulation (GPSR) is a set of rules that requires consumer products to be safe and our obligations to ensure this.

If you have any concerns about our products, you can contact us on ProductSafety@springernature.com

In case Publisher is established outside the EU, the EU authorized representative is:

Springer Nature Customer Service Center GmbH
Europaplatz 3
69115 Heidelberg, Germany

Batch number: 10761599

Printed by Printforce, the Netherlands